by Janice Reynolds

A Practical Guide to @CRM

Building More Profitable Customer Relationships

A PRACTICAL GUIDE TO CRM

copyright © 2002 Janice Reynolds

Published by CMP Books
An Imprint of CMP Media Inc.
12 West 21 Street
New York, NY 10010

ISBN 1-57820-102-0

For individual orders, and for information on special discounts for quantity orders, please contact:

CMP Books
6600 Silacci Way
Gilroy, CA 95020
Tel: 800-500-6875 or 408-848-3854
Fax: 408-848-5784
Email: cmp@rushorder.com

Distributed to the book trade in the U.S. and Canada by
Publishers Group West
1700 Fourth St., Berkeley, CA 94710

Manufactured in the United States of America

Table of Contents

Acknowledgments
Preface

Section I: Determining the Need for CRM
Chapter 1: What is CRM?   001
    A Different Corporate Culture   002
        The Objectives   003
        The Customer   005
    The Internet's Influence   006
    CRM Architecture   006
    The One-to-One Philosophy   008
    In Summary   011
Chapter 2: The Evaluation Process   013
    The CRM Team   013
    Determine the Objective(s)   014
    Perform a Needs Analysis   015
        The Customer's Perspective   016
    The Prospects of Success   017
    What's the Best Strategy   019
    Cost Considerations   020
    In Summary   022
Chapter 3: Preparing a Business Case   023
    The Construction Process   025
        Costs   026
        Return on Investment   026
        Measure the Benefits   028
        Discuss the Risks   029
    In Summary   030

Section II: The Strategy Process
Chapter 4: Creating the CRM Strategy   031
    Where to Begin   033
        Finding a Strategy that Fits   034
    Understanding the Technology   035
        Managing Data   037
        Features and Functionality   039

The Road to Success	042
Actionable Planning	043
Delivering Customer-centric Functionality	045
Avoiding Potential Land Mines	045
Key Strategy Issues	046
In Summary	047
Chapter 5: A Program of Projects	049
Project Management	052
The Project Manager	052
The Project Plan	053
Program Management	054
The Program Manager Skillset	054
The Program Manager's Responsibilities	055
The Program Management Plan	056
Corporate Oversight and Control	057
In Summary	057
Chapter 6: The People Factor	059
Managing the Changes	060
Management Buy-in	061
Communicate	062
Training, and More Training	062
A "Change Management" Plan	063
Spreading the Message	064
Change Impact and Adoption Strategy	064
Employee Buy-in	065
Collaboration	065
Acceleration	066
In Summary	066
Chapter 7: Increasing Customer Loyalty	067
The Reality	068
The Problems	069
Differential Difficulties	069
Customer Service	069
Inadequate Customer Tracking	070
Wrong Customer Focus	070
The Fix	071
A Customer Loyalty Program	071
Create a Customer-focused Environment	072
Institute a Cultural Change	073
The People Factor	074
Benefits	075
In Summary	076
Chapter 8: Know Thy Customer	077
Assessing the Customer Base	079

Identity   079

Loyalty   080

Value Quotient   081

Satisfaction   081

The Results   081

Customer's Roles   081

The Prospect   082

The Account   082

The Business Partner   083

The E-customer   083

The Company Asset   083

The Customer Experience   084

Using Data Wisely   085

The Customer's Perspective   086

In Summary   087

Section III: Technology

Chapter 9: The Many Flavors of CRM – An Overview   089

The History   090

Current Events   091

The Choices   092

Operational   094

Analytical   095

Collaborative   095

Vertical Options   096

Technology in Evolution   096

Software Selection   097

Hosted Solution   098

In Summary   099

Chapter 10: Marketing   101

CRM to the Rescue   101

How It Works   102

Coordination is Key   104

Marketing Automation   105

The Sales Connection   106

Database Marketing   107

Direct Marketing   108

Interactive Direct Marketing   109

E-marketing   110

Closed-loop Systems   113

Finding the Right Fit   113

Something Old, Something New   115

Benefits   115

In Summary   116

Chapter 11: Sales   119
   Planning   120
      The Customer Life Cycle   121
   Sales Automation Choices   123
      The Communications Infrastructure 124
      Tools for the Sales Professional 124
      Automating the Sales Manager   131
   Vertical Solutions   135
   The Human Element   135
      Motivate Don't Mandate   136
      All Employees Aren't Equal   137
   Benefits   137
   In Summary   138
Chapter 12: The Call Center Evolution   139
   The Stampede to CRM   139
      The Bounce Syndrome   140
      "At Risk" Customers   141
      Improving the Customer Experience 142
   CRM and Traditional Call Center Architecture   143
      The Time is Right   143
      Metrics Win the Day   144
   Haste Makes Waste   144
   Evaluating CRM Systems   145
      A Blended Center   146
      Cross- and Up-selling   148
      Data Management, Reporting and Analysis 148
      Computer Telephony Integration (CTI) 152
      Multiple Channel Capabilities   154
      Document Generators   158
      Campaign Management   158
      Knowledge Management   159
      Workflow Management   160
      Call Center Management   160
      Other Considerations   161
   Field Service   161
      Minimum Criteria   162
      Wireless   162
      Advantages   164
      Integration   164
   Call Center Staffing   165
   Costs and Benefits   165
   In Summary   167
Chapter 13: The Importance of Data   169
   The Customer Data Architecture   172

The Data Storehouse 174
   A New Data Architecture 175
   MetaData 176
   Data Cleansing 176
Data Access and Analysis Tools 179
Data Mining Tools 179
The Web 183
Managing Data in the Mid-sized Business 183
Knowledge Management 184
   Customer Intelligence 185
In Summary 185
Chapter 14: Hosted Solutions 187
What is a Hosted Solution? 187
Acronym Alphabet Soup 188
Why use a Hosted Solution? 190
   Costs 192
   Implementation 192
   Limitations 193
   Advantages 193
Differences between Hosted Solution Models 193
   Application Specific 193
   Tailored Solution 194
   Vendor Specific 194
   Menu of Solutions 194
   Vertical Solution 195
   End-to-End Solution 195
   Business Solution Provider 195
Benefits 196
   Time-to-Benefit 196
   IT Skillsets 196
   The Sweet Spot 197
   Web 197
The Concerns 197
Trade-offs 198
The Decision Process 199
   Due Diligence 199
Integration Woes 202
   Time 203
Costs 203
In Summary 204
Chapter 15: Partnering for Success 205
How to Find a Tech Partner 206
Consultants 206
   Why Hire a Consultant? 207

What to Avoid?	209
Consultants' Wrap-up	210
Systems Integrators	210
How to Find the Best Fit	212
Advice from the "Horse's" Mouth	213
What to Expect?	214
Systems Integrators' Wrap-up	215
The Selection Process	215
Determine Need	215
Qualifications	216
The RFP	216
The Interview	217
Due Diligence	218
The Contract	218
Work Plan	219
In Summary	219
Chapter 16: The Vendor Selection Process	221
Understand the Vendor Community	222
Verticals	223
Vendor Selection	224
Selection Criteria	225
Define the Need	225
Narrow the Field	226
The Short-List	227
Request for Proposal	229
RFP Committee	230
Criteria	230
An Example RFP	231
Pre- and Post-RFP Tasks	237
Legalese	237
And the Winner Is...	237
Costs	238
Vendor's Vision	238
Due Diligence	239
Negotiations	240
Paperwork	240
In Summary	241
Chapter 17: Implementation and Deployment	243
Implementation Philosophy	244
The Implementation and Deployment Plan (I&D Plan)	244
The Implementation Team	245
The Systems Migration Plan	245
Process Re-engineering	247
Moving Forward	251

Finalizing the I&D Plan	251
Middleware	254
CRM Needs Middleware	255
Enterprise Application Integration (EAI)	257
Avoiding the Pitfalls	257
Front-office/Back-office	258
The Channel Conundrum	258
An Amalgam	259
Testing	259
Security	260
CRM Ups the Security Risk	260
Final Deployment	261
Training	261
Post-Implementation	262
Post-Implementation Team	262
Documentation	263
Support	263
Maintenance	264
Costs	264
In Summary	265
Some Parting Advice	267
Glossary	269
Index of Figures	287

Acknowledgments

My books are always somewhat of a collaborative efforts — friends drag me away from my computer to remind me that life exists outside the publishing industry, my editor makes me sound intelligent and the CMP people (Christine Kern, Matt Kelsey and Lisa Giaquinto) offer me lots of encouragement.

I would especially like to thank Robbie Alterio, whose talented design and layout skills give the book a smart milieu.

Preface

Customer Relationship Management (CRM) has its roots in the age-old principle of "the customer is always right." But CRM is much more than that, since it identifies how to profitably act on that axiom — at all times and across all channels and functions. Moreover, even in its simplest form, CRM defines the way a company FINDS, GETS, and KEEPS its customers."

Why are so many within the business community jumping on the CRM band-wagon? One could list a number of possible reasons relating to socio-cultural factors, business and technical advances, but the real impetus for CRM today is its promise of competitive differentiation in a parity environment.

Today's corporations find it increasingly difficult, if not impossible, to compete on the basis of product alone. Technological advancements have enabled the near-immediate replication of product features and functions by competitors. These days just a few weeks stands between a new product launch and a saturation of the market with similar, "reverse-engineered" products made by competing manufacturers.

Can you think of many products that are truly unique? I can't.

Price, which has traditionally been a basis of competitive differentiation, is no longer a means for many companies to compete. Complex channel networks have caused parity pricing. Promotional strategies no longer provide the necessary means for differentiation — "buying clubs" abound, special offers and sales are the norm.

The adoption of a CRM strategy can provide a more meaningful sales and service experience. It can give customers a reason to frequent your business rather than that of your competitors. It can be a means of differentiation.

Businesses that adopt CRM recognize that their customers make buying decisions based on their overall experience — an experience that not only includes product and price, but also sales, service, recognition and support. If companies can get all of those factors right, on a consistent basis, they will be rewarded with ongoing customer loyalty and value, which results in a larger profit margin.

One of the compelling issues that drove me to write this book is that many in the business community are struggling with their CRM projects. This is mainly because the companies err in believing that they can achieve a utopian CRM scenario through the implementation of technology alone. They can't. CRM can add significant benefits to a business's bottom line only when a company turns itself inside out by adopting a customer-centric business strategy which drives a customer-centric corporate culture, then brings onboard technology which is properly implemented and integrated into the company's IT infrastructure. Any company that thinks it can just "plug it in, turn it on" and then be rewarded with stellar customer relationships is in for a rude awakening.

A further cause of confusion when a company starts down the CRM path is that no one vendor provides all of the CRM capabilities needed for a corporate-wide CRM initiative. Furthermore, even the depth and breadth of functionality provided within a specific area will vary from one vendor to another.

The CRM vendor marketplace is a free-for-all. This book is an attempt to help the reader wade through the quagmire of CRM vendors' offerings. Some CRM tools have been developed to provide the increasing functionality needed by sales and marketing departments. Other tools have been imported from various other applications to round out a CRM product suite. Still others are developed to satisfy the needs of the new customer-centric (vs. product-centric) business philosophy. Industry pundits are having a field day concocting names and acronyms for these CRM-related components, which only adds to the confusion.

What do I want you, the Reader, to come away with as a result of perusing this book? This can be summarized as follows:

CRM is not a technology initiative. Many consider CRM a technology and thus assign their CRM implementation project to the IT department. CRM conferences are often equated with technology exhibits and demonstrations. Admittedly, technology is needed in order to implement CRM, but it's not the driver of CRM nor the solution.

CRM is not exclusively a marketing initiative. Many have likened CRM with customer-focused marketing or database marketing. While employing CRM in a marketing department can result in more effective, data-driven marketing efforts, this alone does not make a total CRM strategy.

CRM is not exclusively a sales initiative. The sales department (or the marketing department) is often the first to clamor for CRM, but sales is just one functional area out of many that can benefit from CRM.

CRM is not exclusively a service initiative. As with sales and marketing, customer service is only one functional aspect of a successful CRM initiative, but customer service is not the sole driver of a CRM strategy.

In actuality, CRM is a strategy that is enabled by technology. There must be a CRM strategy in place that envisions and blueprints a corporate-wide CRM initiative involving marketing, sales, service, data storehouses and technology, as well as the other inner-workings (people, culture, environment) of a company.

That's not to say that a CRM strategy can't be implemented one department or one point solution at a time. It can and, many times, should be. But it must be done within a framework that encompasses a corporate-wide CRM initiative that includes all employees, all work and business processes, all systems and all data storehouses within the company. There should be in place an overall design or blueprint that maps out how every person and system within the corporate walls shall eventually work together in harmony toward the common goal of stronger and better customer relationships.

Having even one broken spoke in the wheel — one area that is less than committed to CRM — can make the difference between success and failure of not only the CRM initiative, but, perhaps, the business itself.

Janice Reynolds
New York, NY
February 26, 2002
janrey@bellatlantic.net

Chapter 1
What is CRM

H ow well a company maintains its relationships with its customers may well determine its success. Knowing its customers and focusing on their needs enables a business to better deploy resources towards an effective outcome for everyone.

Since the first cave man profitably traded a bundle of furs for flint to make arrowheads, business strategies have focused on cultivation of the customer. Thus began the millennia-long march to refine techniques in customer relations. Over the centuries, many improvements were made to routine and manners, but not to strategy and applications — that is until late in the 20th century.

With the debut of affordable, computerized technology in the late 1980's the first generation of CRM tools appeared. These DOS- or Unix-based products, usually referred to as "sales force automation" solutions, were, in reality, not much more than contact management tools. These products embodied little or no strategy or customization. It's only with the convergence of information technology, telecommunications, and the Internet that the whole concept of CRM really comes alive.

In the early to mid-1990s, e-commerce entered the scene offering customers more control over their purchasing choices, albeit at the sacrifice of personalized service. As customers exercised this power, companies looked for innovative ways to bring their increasingly fickle customers back into the fold. One early approach was to analyze shopping patterns. Soon software vendors of every shape and size jumped on the CRM bandwagon, offering tools that could automate sales and marketing processes, as well as enable call centers, dispatch, and other customer-centric business process. However, once again, technology, not strategy, drove the changes.

This second generation of software (which some pundits referred to as "opportunity management systems" or OMS) was generally Windows-based, and usually offered client-server support for relational databases, such as Informix,

Oracle and Sybase. These too, offered little in the way of customization opportunities. Yet, they could normally interface to back-end systems allowing data transfer to and from legacy databases.

By the end of the 20th Century it was possible to find what is know known as "customer relationship management" (CRM), solutions. These new tools are the technology end of CRM and they can automate all processes providing information relating to customer interaction so it's available corporate-wide.

The business and technical communities finally got it right — CRM is a *strategy* that is enabled by *technology* (products, tools, software, etc.). To this end, CRM tools enable the integration of information management systems so that a wide variety of data spread throughout a far flung enterprise can be used to plan, schedule and control pre-sales and post-sales activities.

Since CRM is not just a technology, but is a new way of doing business, the comprehensive definition of CRM might be "the business strategy, process, culture and technology that enables a company to optimize revenue and increase value through understanding and satisfying the individual customer's needs."

GartnerGroup, a premier research and consulting firm, also attempts to define CRM. Their definition works within the context of this book as well:

WHAT'S MEANT BY CRM?
CRM is a management discipline — a philosophy even — that requires businesses to recognize and nurture their relationships with customers. CRM melds customer intimacy with economies of scale. It enables the building of close relationships between the representatives of a business and its customers.

With CRM, customers' needs and preferences are available to anyone in the business working at the customer interface, and customers are treated consistently regardless of channel. Such interactions can feel like one-to-one relationships.

By putting the customer at the heart of the business, CRM is a manifestation of customer focus — a principle that's widely espoused but rarely delivered.

CRM is essential when a key strategic thrust of a business is customer intimacy. The idea of CRM is not new, though. Local, small-town stores provide a CRM-like service, anticipating the needs of customers based on intimate knowledge of their circumstances and preferences and treating different customers in different ways.

✪ A DIFFERENT CORPORATE CULTURE

CRM brings new ways of doing business. It's a corporate culture change wherein the business focuses on the customer rather than the product. The biggest problem with CRM isn't software, it's a lack of strategy — codifying "customer first" corporate values, practices and processes.

For CRM to be successful within the corporate environment, the executives and management *must* view CRM as an opportunity to shift the business's focus on an emphasis to serving the *individual* customer. This requires a critical review of all business process and a re-orientation of the corporate culture. Every employee, from the president and department heads to the customer service representatives, invoice clerks and warehousemen, needs to realize that they have a direct hand in building the right relationship with each and every customer. This means the entire corporation must buy into the CRM culture and learn to think in a customer-oriented manner.

Before the advent of the Internet, customers expected to reach a company via a personal visit, telephone, fax or regular mail. Today, customers expect to contact a business via a variety of channels, which can include not only the aforementioned, but also email, websites, ATMs or kiosks, and even video. Although the Internet does allow a company to reach more customers than ever before, it also allows any competitor to reach out and lure established customers away, if a company isn't attentive enough. CRM solutions allows a company to improve the quality, consistency and reliability of its customers' experience with each and every interaction, which enables a company to guard against competitive erosion while acquiring and retaining new customers.

Many within the business community still don't understand that CRM isn't just technology. Thus, businesses often approach their CRM projects bass-ackwards —as a means of linking the different customer-facing functions within a company. (The technology.)

To implement a CRM initiative correctly, the entire company must change the way it thinks about its customers: Who are they? What do they want? How can they be better served? (The objectives.) CRM is a way of changing a company's mindset and, therefore, the first step is to overhaul the company's business plan. (The strategy.) Technology comes later — to enable the strategy so as to reach the business's CRM objectives.

The Objectives

The first step for a business should be to determine its most important CRM *objective*. Should it sell more product to existing customers? Increase sales by extending its customer base? Or work on keeping the customers it already has through better customer service and loyalty programs?

If the objective is customer retention, then perhaps the business should prioritize customer service improvements, such as implementing a web-based self-service system. If customer acquisition is the main objective, the company may want to focus on sales and marketing process improvements. There's a whole laundry list of business objectives that CRM can help a company to achieve, for example it can:

* Help sales staff to close deals more quickly and more efficiently.
* Simplify processes within the marketing and sales departments.
* Help discover new customers.
* Provide better customer service, which can include improving the efficiency of a business's call and help centers.
* More effectively cross-sell and up-sell products and services.
* Increase revenues through improved customer relationship management.

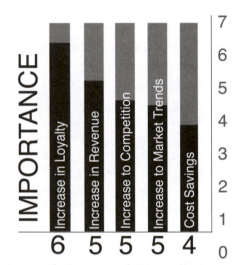

Figure 1. The Importance of Business Objectives in the Deployment of CRM Initiatives. Note: Responses are ranked on a scale of 1 to 7 in which 1 is the lowest level of importance and 7 is the highest. Source Gartner Dataquest (May 2001).

A business should perform a thorough analysis of its markets, and customers' needs, distribution channels, and current technology to determine what CRM functions are required. For example, a rubber band manufacturer doesn't need sales configuration tools, but a computer company just might. A company that sells and services turbine engines might need a CRM solution that can help it manage field service and dispatch, although a financial institution doesn't need such tools. A business that sells telecommunications equipment will have ample opportunity to use contact and opportunity management software, but the local installer probably doesn't need such complex technology.

To accomplish any stated CRM objective(s) requires some type of CRM software package, i.e., a "cooperative" of CRM tools working in unison to bring to fruition a company's CRM strategy. CRM isn't cut and dried, it wears many hats and cuts across the company's business process. A business, for instance, may want its CRM system to include any, or all, of the following functions to obtain it's stated CRM objective(s):

* Contact and opportunity management.
* Sales force automation.
* Sales configuration.
* Field service and dispatch.
* Marketing campaign management.
* Call center management.

All of these technologies (whether sold via product suites, as an individual software package, or whatever) fall under the CRM umbrella. However, it's the strategy that ultimately turns these individual technical elements into a productive and successful CRM initiative.

The Customer

But, the reader might ask, "How does CRM accomplish all of this?" Not by simply buying and installing software. For CRM to be truly effective it first must be determined what kind of customer the company needs to acquire and retain; then what information should be gathered, and then how to utilize that information to accomplish its CRM objectives. For example, many insurance companies keep track of their customers' life stages in order to market appropriate products, such as term or whole life insurance to the young family; annuity policies to the empty nesters; automobile and homeowner insurance at just about any life stage.

Next, identify the many different ways customer information is currently obtained, where and how this information is stored, and how it is being used. A business may interact with its customers in myriad ways, including marketing and sales campaigns, websites, retail locations, call centers, and an outside sales force. A company-wide CRM system would link up each of these touch points, collecting data as it flows between operational systems (i.e. sales and inventory systems), and analytical systems (e.g. personalization systems) that can help sort through the data for patterns.

Company analysts could then comb through this data to obtain a holistic view of each customer and pinpoint areas where better service is needed. Using the insurance example again, if one identifiable customer has a personal life insurance policy, a key person policy, business interruption insurance, and a large mortgage insurance policy, it behooves the insurance company to provide that customer with preferential treatment each and every time there is contact.

Often CRM is described as requiring systems that provide a single view of the customer across all touch points and all channels. But that's not necessarily true for all companies. Not all businesses or customers are equal. For example, a business may want to concentrate on identifying its customers and implementing *cost-efficient* ways to keep them, especially the profitable customer. While an integrated CRM system is the most potent way to accomplish such objectives, a point solution also can provide most of the same benefits. For example:

CRM software that allows a field sales force to link to the company's website via a laptop or handheld computer (via either a traditional telephone connection, DSL, cable or even wireless), can offer the field sales staff access to a customer's sales history in real-time. This can help the sales staff finalize a sale.

In a call center, CRM can enable the center's agents to view a customer's sales and payment histories, repair and order status. There's even tools that allow customers and agents to surf the company's website together while conversing about the customer's issues.

⊙ THE INTERNET'S INFLUENCE

It's impossible to discuss CRM without bringing to the fore the exponential growth of the Internet and its subset, the World Wide Web. The tremendous growth of the online community (business- and consumer-based) changed how a company manages its customers and, thus, changes the very definition of customer relationship management.

Due to the influence of the Internet, businesses increasingly find that their customers are demanding the ability to interact with them in real-time. Yet, because the traditional customer relationship management systems are meant for use through independent channels, many a company stumbles in its delivery of a consistent customer experience across all of its channels — from pre-sale to post-sale support.

Today's CRM tools provide linkages and integration of data from and through all customer "touch points." The businesses that have integrated this new breed of CRM tools now have a competitive advantage. Take the example of a company that has linked its website to its call center. Now it can track a customer's sales history and buying habits. When a customer logs onto its website to buy an item or service and encounters a problem, an agent in the call center can simultaneously "talk" to the customer, view the customer's sales history, and devise a sales pitch for a new service ad hoc. The interaction may also yield valuable information about buying habits, which, if stored and analyzed correctly, could affect future products or services. The results: the company has a satisfied customer, made a sale, and may have learned a bit more about its customer in the process.

⊙ CRM ARCHITECTURE

A full-spectrum CRM architecture will provide integrated automation of business process that encompasses customer touch points. These normally include sales (for instance, contact management and product configuration), marketing (this might include campaign management, telemarketing, data mining), customer service (this is where call center, field service, email management come into play), and all data storehouses.

The architecture will also accommodate a number of customer channels. Customer communication streams in through a variety of channels: in person, telephone, chat, email, fax, and more. The customer experience must be identical no matter what channel is used.

Next, the CRM architecture will integrate the customer information that flows through these diverse departmental touch points and customer channels. The technical infrastructure must provide a way to process this mountain of information (which is sitting in various data storehouses) so it's available when and where it's needed.

The marketing gurus separated CRM into three distinct areas to help everyone to better understand how a CRM strategy and technology can provide such an all-inclusive architecture that's focused on serving the customer.

1. Collaborative CRM. The collaborative interfaces (e-mail, conferencing, chat, real-time applications) that facilitate the interaction between a company and its customers, as well as within the business itself when dealing with customer information (e.g. customer service to sales, sales to marketing).

2. Operational CRM. The automation of integrated business process that involves front-office customer touch points (sales, marketing, and customer service) via multiple, interconnected delivery channels, including the integration between front- and back-office (i.e. sales automatically pulling data out of inventory systems or customer service automatically calling up a customer's billing records).

3. Analytical CRM. The analysis of data for the purpose of business performance management. It refers to the analysis, modeling and evaluating of data, which is sitting in a data warehouse, a data mart, or various data storehouses to create a mutually beneficial relationship between a company and its customer.

Many businesses will soon move to WAP-based (or other wireless) devices, especially for field service, dispatching and field sales. Again, the engineers are ready; many CRM companies offer a version of their CRM tools for the wireless crowd's use. Getting the architecture right for the wireless environment is a tough challenge, so be wary and keep an open mind when looking at these specific CRM offerings.

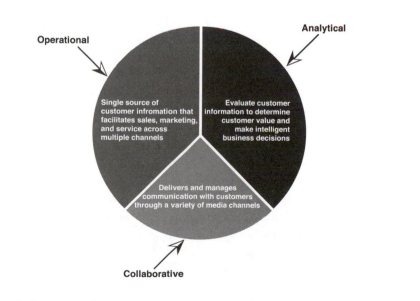

Figure 2. Components of CRM.

✪ THE ONE-TO-ONE PHILOSOPHY

The business world's fascination with products and product differentiation has been replaced by a new fascination — customers and customer differentiation — leading to the rebirth of the one-to-one philosophy.

A one-to-one philosophy might be described as customer-centric with the focus on the individual customer. Thus, a company builds a one-to-one relationship with each customer, and retains and grows that relationship in a profitable and productive manner.

In an increasingly competitive business environment, companies struggle to attract and retain customers. Though the current crop of CRM tools is relatively new, their innovators have given these products the means to increase the value of a business's established customer base through the ability to develop a "learning relationship" with the individual customer.

One-to-one marketing, relationship marketing, and even database marketing are used interchangeably to refer to efforts to accommodate the customer, based on knowledge and information unique to that customer, i.e. one-to-one CRM. Whatever it's called, this kind of marketing becomes increasingly important as businesses search for ways to understand and utilize the myriad ways they can interact with their customers.

Many CRM products address, or attempt to address, this type of marketing. But for any of them to succeed, the business community must first conquer two major obstacles: their legacy systems and the lack of a centralized customer database.

Unless a company can find ways to link its legacy systems and bring all of the information in its diverse databases into one central data depot, it cannot implement a system that can tell the company, at large, whether a specific customer is high-value, low-value or even at risk because company personnel can't access the required data when needed. A business can't discern between the customer who has bought $100,000 worth of product over the past year, a kid who's looking for a bit of help for his or her homework assignment, or a valuable customer who's at risk because of a previous hitch that caused them grief.

To put the one-to-one philosophy into practice requires a business to identify the 20% of its customers who bring in 80% of their life time value. Then the business must be able to retain that 20% — even increasing its share of that group's volume of trade. This requires an enterprise suite of CRM components, i.e. products from Siebel, SAP, Vantive and the like.

Even with a less intensive point solution approach, CRM is critical to a business's success in today's environment and is central to any effort a business may make to establish a lasting, long-term, profitable customer relationship. A CRM system can definitely help level the playing field — if done right.

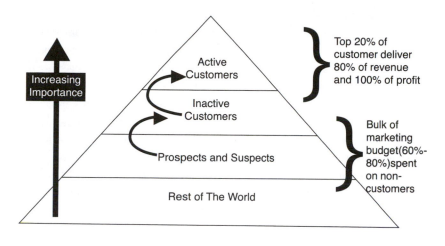

Figure 3. The Customer Pyramid.

The One-to-One Relationship

The term most often associated with CRM is one-to-one marketing, which refers to marketing techniques aimed toward creating a one-to-one relationship with the customer. I've written this section to give the reader a primer into the one-to-one philosophy as it relates to CRM, since any CRM initiative in one way or another must adopt *some* elements of the one-to-one philosophy.

With the one-to-one relationship model at the top, other CRM initiatives, such as revenue-generation through cross-sell, up-selling and add-on opportunities are simply a process of anticipating a customer's needs and then offering to meet those needs. An example: A customer who normally orders 10 reams of paper a month but doesn't order printer or copier cartridges. A special aimed at that customer has a good chance of capturing his or her printer and copier cartridge business.

Price soon becomes secondary to the customer once there's an established one-to-one relationship. Not that price won't be a consideration, it will, but it won't be the primary issue because the customer will develop a vested interest in continuing the business relationship. Why? The customer knows that the company is price-competitive *and* knows that the company is familiar with his or her needs and wants. The company doesn't waste the customer's time. As long as the company provides the customer with the goods and services that fit their individual requirements, there is no need for them to explore alternatives.

A Management Discipline

Adoption of a one-to-one philosophy requires more than changing marketing and sales plans. It's a management discipline that puts the customer at the heart of the business. And it's a discipline that applies to almost any kind of customer — the business customer as well as the individual consumer. It's based on a collaboration principle, engaging customers in dialogue, building trust so that they share their needs and aspirations, their problems and ideas. This requires a long-term view:

* Information gathering — not only name, address, credit card information, but also purchasing history, desires (wish lists), and so forth.
* Deployment of personalized fulfillment — the employees know (for example) that the customer always requests a specific third party address on mother's day or for all products to be delivered to a Mail Box Are Us location.
* Closed loop feedback — as information from customers makes its way back into the business's systems, the data generates new actions and process — based on these continuous streams of data. This not only helps to improve sales and marketing activities and customer relationships, but also provides a way to measure marketing ROI.

Obvious Benefits

Implementation of a one-to-one philosophy provides obvious benefits:

* The business focus shifts to the most profitable customers.
* The lifetime share of the customer base, not market share, becomes a key performance indicator.
* Keeping profitable customers is cost-efficient and far outweighs the cost necessary to acquire new customers.
* Better bang for the buck — products and services are more accurately targeted due to in-depth knowledge of the individual customer.
* Customers, as collaborators, become a rich source of ideas for new product development, leading to high post-launch success rate.
* Comparative shopping among the customer base becomes a thing of the past. How? Easy — mass customized offerings that cater to specific customer groups within the customer base.

The Principles of Success

Take an active interest in understanding the valuation of the customer base as a primary asset and then treat it as such, while carefully managing the investment made in this asset. Allocate the company's management and employee time and effort to those customers who will yield the highest ROI.

Customize cross-selling, up-selling and add-on opportunity to the individual customer's needs and preferences. Take advantage of the feedback loop with an individual customer — it's useful only if the interaction from the customer can be and is incorporated into the way the company behaves toward that customer. Now a company can expand its product offerings to increase value at three levels: the core product (customer bought Zip drive, offer Zip products), product-service bundle (customer makes extensive use of technical assistance, offer service policy), and enhanced-need set (the reams of paper/ink cartridge scenario).

Communication and feedback from individual customers drive the one-to-one relationship. However, before a business can solicit feedback, it must create a dialog. This is easily accomplished if all forms of customer interaction are integrated and consistent. Think of customer interaction and dialog as revenue-generating activities, not as costs.

The creation of a one-to-one relationship management system requires not only an effective CRM strategy and enabling tools, but also a fundamental change in the business's management philosophy, design, and support systems. A key requirement to making the transition is the business's ability to link customer data across functional and divisional boundaries; but sharing customer data on a corporate-wide basis requires a high degree of inter-company communication and data discipline.

CAVEAT: *Don't let the allure of a one-to-one model entice the company from its CRM objectives. Always keep in mind that the best CRM solution isn't always the most advanced technology, but instead it's the one that allows the company to meet its stated CRM objectives.*

Many of today's businesses are far-flung — even many family owned companies have more than one location. Consequently, in many businesses the familiar one-to-one philosophy has been replaced by the any-to-any philosophy. No longer does a customer walk in the door (or telephones) a business and the staff automatically knows him or her by name. In today's business environment, the staff typically can't identify the "special" customers. In days gone by, when those "certain" customers walked through the door, the clerks' eyes actually lit up — they put on that extra smile, and interjected a lilt in their voice — the clerks "just knew" who were the "big spenders". Now any staff member might be the next to serve a customer, there isn't any differentiation between the way that customer or the next customer to walk in the door is treated. CRM may be able to provide the level of customer intimacy that has become lost due to the scale of business today.

✪ IN SUMMARY

The key to any business's future success is its existing profitable customer base. This is accepted marketing wisdom, but, sadly, it's one to which many companies just pay lip service. Here is how serious the problem has become: Study after study show that close to one-half of the businesses in operation today cannot identify the principal causes of customer churn (turnover); even after the customer has been lost. Out of the small majority that state they can identify the reasons for customer turnover in their company, only a paltry 20% or so could specify how the process worked. But, even in that small group there was a noticeable inability to predict when a customer was at risk. If a company can't identify the customers at risk (which almost the entire business community admits it can't), what's to keep these companies from losing their most profitable customers?

Chapter 2
The Evaluation Process

I f ten executives within the company were asked to describe their CRM goals, there would be ten different answers. This is not to say that what each executive expects their department to gain from a CRM initiative is incorrect, rather, just as in life, relationships are complex. This is why an evaluation process is so important.

Many CRM blunders are due to a failure to articulate a company's needs and, thus, the company buys features and functionality it doesn't need. Its staff can end up struggling with an overly complex application, a system that requires massive customization to fit the company's needs and infrastructure, or a system that is inadequate. With the proper evaluation and strategy, these quagmires can be avoided.

✪ THE CRM TEAM

How to find the right CRM solution? First, create a strong internal team (the "CRM team") to evaluate, build and continually update the company's CRM objectives and strategy. At the outset, the CRM team should consist of the CRM evangelist, an executive sponsor, IT management, and management from each department that will be touched by the CRM initiative. These individuals will be the sponsors within their departments for the changes that must be wrought to bring about a successful CRM initiative. A project manager (and/or program manager) and others will be added as the CRM strategy and initiative evolves.

Note: The CRM team is like a football team, which consists of a set number of players, coaches and support staff but the players that are on the field will change as the situation dictates. The same with the CRM team. It will consist of a set number of members, which is usually members of the IT staff, executive sponsor and CRM evangelist. Sooner, rather than later, a project and/or program manager will join the team for the long hall. Others will come and go as the need dictates.

The CRM team will:

* Provide a detailed road map of the project before beginning the implementation process, i.e. basic project management details.
* Obtain executive level buy-in. This is why it's vital that any CRM project has a strong advocate on the company's executive board as a CRM team member.
* Develop a budget and timelines. Note that the CRM initiative (just as with any IT project) will run into unforeseen problems, so allow for the unexpected.
* Address cultural objections. Problems can arise, especially in departments such as sales and marketing.
* Emphasize the importance of training. Failure to invest in training (and refresher courses) can greatly lower the probability of success for any CRM initiative.
* Continuously evaluate and consider additions to and upgrades of the CRM initiative as the business's needs evolve over time.

◯ DETERMINE THE OBJECTIVE(S)

The CRM team can establish the company's key business requirements by listing the features and functions most important to the company's business and infrastructure.

For instance if *customer acquisition* is the main objective of a CRM project, the company needs a CRM solution that understands what, when and where the needs of its customer base lie. This is the bailiwick of sales force automation (SFA) tools and campaign management solutions. But, also look for tools that can provide (a) insight into why prospects choose not to do business with the company and (b) have the ability to forecast the changing needs and expectations of the company's customer base. The best way to accomplish this is to invest in developing appropriate customer-facing interfaces.

One example would be a web-based CRM solution, which allows a company to listen to its customers through surveys and online feedback. By listening to its customers, the company takes the first step towards understanding their needs. There's an added bonus — the analysis of web usage habits. This data stream can provide valuable information that can help the company to reduce customer churn, provide cost effective customer service, and *retain* its customer base. In addition, a top-notch website can help a company to increase brand awareness, provide a comprehensive knowledge base for customer self-service, and provide up-selling and cross-selling opportunities (*customer enhancement]*.

If the primary business objective is *customer retention*, then the CRM initiative should center on continuing to meet the needs and expectations of the company's existing customer base. Businesses that learn how to please their customers by listening to them, and then respond through the further development of products and services that fulfill their needs, will ultimately celebrate improved profit margins due to increased customer retention rates. While a 5% reduction in customer churn may seem minuscule, that 5% can have a significantly bearing on a company's profits.

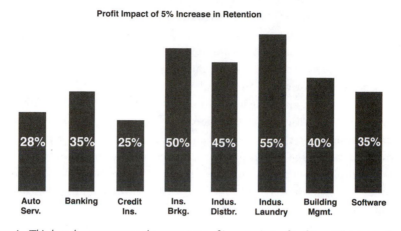

Profit Impact of 5% Increase in Retention

Auto Serv.	Banking	Credit Ins.	Ins. Brkg.	Indus. Distbr.	Indus. Laundry	Building Mgmt.	Software
28%	35%	25%	50%	45%	55%	40%	35%

Figure 4. This bar chart represents the percentage of increase in profit after a 5% increase in customer retention rates.

By targeting desired customers, a company can increase its per customer profit ratio. This requires CRM components that enable a company to understand the worth of each customer or customer group, perform economic value analysis, and other customer valuation methodologies. But, if applied correctly, a company will see an increase in customer profitability.

Effective cross- and up-selling of related products and services to existing customers is referred to as *"customer enhancement."* Customer enhancement is dependent upon the interconnectedness of shared customer information, which provides the ability for a company to interact with its customers, at each customer touch point. This presents the potential to sway a customer's decision to purchase additional services and products. Without such interconnectedness a business can't quickly and efficiently understand how to give its customers what they want — where, when and how they want it.

✪ PERFORM A NEEDS ANALYSIS

The only way to determine if and how the company's current ecosystem can support the requirements of the CRM initiative is to perform a needs analysis. IT personnel should be involved. The technical infrastructure plays a major role in any assessment of costs, by having IT personnel onboard, the CRM team can obtain more accurate budgets and tighter control of the costs of the CRM initiative.

First, decide whether the company's general direction will support CRM or if a shift in focus will be necessary.

Second, analyze how customers proactively and reactively interact with the company. This analysis will help everyone to understand how the business reaches out and touches its customers as well as to identify the contact channels used by individual customers. Also look at the database marketing processes currently in place. This analysis enables the CRM team to understand the consistency or inconsistency of customer treatment across all channels — how different customer segments are treated, and how current initiatives are coordinated between different channels.

Third, examine the business's core business and work processes. Evaluate all current process for fit within the overall CRM strategy. Identify areas that need immediate attention and note what must be re-engineered or replaced prior to integration of any CRM tools. (If leaders from each impacted area are brought onboard at this point, the CRM team can ease resistance to change and heighten the overall awareness of the proposed CRM initiative within that area.)

Look for places where customer information or customer differentiation can add value, such as inventory, customer service, or customer billing. For instance, Dell and Saturn both know what customer belongs to which product as the product traverses the company processes from initial sale to the assembly line through post sale procedures.

Fourth, define what data is critical for the support of a CRM initiative and then determine what approach and tools will be used to analyze this data to make it meaningful and accessible. Terabytes of customer data are distributed across multiple systems (some in departmental silos). Effective CRM is dependent upon data — but the data must be analyzed and interpreted, then cleansed and collated into a central data repository before it can provide usable information for a CRM initiative.

Fifth, consider the existing technology. This means both the front- and back-office systems. Here it is vital that the CRM team understands the level of integration that exists and the changes that will be needed to support the CRM initiative. Then the CRM team must determine the type of technology (CRM components, supporting software and middleware, and hardware) required for the CRM initiative to be successful.

Finally there's the human element. Central to any successful CRM initiative are the end users since they will be using the system on a daily basis. Query these users (the company's employees) and get their wish list.

The Customer's Perspective
Understanding the company from the customer's point of view is an important step toward implementing a successful CRM initiative. Do a needs analysis from the customer's perspective. Many times a customer's experience is inconsistent due to the numerous channels available — retail outlets, telephone, fax, email, and the web. For instance, a customer contacts customer support with a technical problem and receives no satisfaction, but later when airing the complaint to his or her sales representative the customer is provided with a simple fix to the problem.

Find out the customer's view of the business — what it's like to buy something from the company. Many times it's an eye-opener. Use the various channels offered to the customer — place orders over the web, telephone customer support, return an ordered item, send an email inquiry about an invoice, complain about a problem by telephone and email, fax a request for more information about a specific product. During this process, use only the contact points touted by the company's marketing department; e.g. don't use "inside" information. If Joe Sales' number isn't available to the average customer — don't use it.

Next, interview the customers to learn about their own buying experiences, document these experiences so it's possible to dissect not only where breakdowns occur, but also where everything seems to run smoothly.

Note: *The more contact channels available to the customer, the greater the need for a CRM initiative that can provide a single centralized customer view.*

Unless a company's numerous customer touch points and channels are properly managed, its ability to leverage the accrued knowledge about its customer base is limited. Understanding the entire customer experience allows a business to see where its systems work and where they breakdown. K-Mart has learned this lesson the hard way.

K-Mart admits that its customers have suffered daily from departmental disconnects. For instance, K-mart's marketing department customarily places sales inserts into regional Sunday papers. But, it's not uncommon for eager customers to be disappointed when they enter a K-Mart store bright and early Monday morning only to find the advertised special to be out of stock.

Although K-Mart readily issues "rain checks," a customer's disappointment often can lead to customer attrition. The unsavory situation at K-Mart arises out of departmental communication failures — between the marketing department, logistics and fulfillment, and regional retail outlets. But it's never too late to learn from your mistakes. K-Mart has embarked on a massive CRM initiative to remedy this situation and other customer-related problems.

✪ THE PROSPECT OF SUCCESS

Now that the CRM team has assessed the need for a CRM initiative, the next step is to determine the prospect of success. Poll the executive management to determine if they will actively support the initiative. Once doubts are brought to light, many problematic issues can be corrected before a business case is presented.

As the CRM team takes on the task of assessing the prospect of a CRM initiative's success, the CRM evangelist should take a feasibility inventory. To illustrate:

* ✻ Does the project have top management commitment? Without such commitment, the authority to make the sweeping changes necessary for a CRM project to succeed is virtually impossible. Continued funding of the project could also be jeopardized.
* ✻ What is the differentiation of the company's product(s) and/or service(s) within its market place? If differentiation is low (as with breakfast cereals), a

CRM system is more appropriate than when differentiation is already high (as with computer operating systems). With low differentiation a CRM system can help a company stand out from the crowd. Industries in which CRM is most likely to aid in product differentiation include financial services, insurance, and consumer durables.

✳ How is the company placed within its market place — is it a start-up or well established and successful? If the business is a start-up, the chances of a CRM initiative being a success are good since there are few ingrained procedures or legacy systems to deal with. If the company is well established, the path is more complex. For instance, redefining business rules and process and data cleansing are examples of some of the more difficult aspects of implementing a CRM system in an established business environment.

✳ Is there a consensus within the company that there is not only a need to cater to a growing range of customer contact channels, but that those channels should be presented in a consistent and harmonious manner to the customer?

The CRM team needs to look at the customer contact points themselves. Today's multichannel customers have exposed the flaws of inconsistent communications, messaging, product offerings, and service from companies that have departmental silos of information — they don't share! This has a direct impact on customer satisfaction, which, in turn, drives customer churn.

It's essential that the CRM team take a long, hard look at the company's call center operation, which *should be* the core of a business's multichannel communication byways. As sales, marketing, and service channels continue to expand, many within the business community have found that their call center is outdated.

The traditional call center has a limited channel focus: telephone, fax, and "snail" mail. When a company initially brings other channels onboard, such as email, web chat and web-based call-back opportunities, it establishes systems (or doesn't establish, as is more generally the case) that are separate from the call center's domain. Many times email information isn't even available to the call center personnel. The remedy is to upgrade the call center facilities to handle both existing and new channels.

Now, define which customers — corporate end users, individual end users, resellers, all customers, etc. — are to be the focus of the CRM initiative. Then, pinpoint how the company can benefit from developing closer relationships with those customers.

Basic CRM Requirements

To sidestep some of the difficulties that can hinder or even block the success of a CRM initiative ensure that the CRM team has determined that:

✳ The CRM initiative has good proactive executive sponsorship.
✳ The company has well-defined objectives, goals and strategies for the CRM initiative.

* There are experienced project and resource management onboard.
* A plan to counter expected resistance to change within the company is in place.
* A plan to deal with mismatched technology infrastructure has been drawn up.
* It has accounted for the impact that the CRM initiative will have on existing business process.
* It has planned effectively for in-house awareness and training.
* The company has achieved the culture shift necessary for a smooth transition from a product-centric model to a customer-centric model.
* It has overcome financial constraints.

CRM automates and enhances the customer-centric business and work processes of departments, such as sales, marketing, and service. But a company can't expect CRM tools to successfully automate these processes without cleansing its data, tweaking its front- and back-office applications, and providing for adequate initiation and training of its personnel. Then to spew forth consistent data there must be cohesive integration of CRM components with front- and back-end systems. The employees' use of that data provides the customer experience that, in turn, gives the desired ROI.

Nonetheless, all is for naught if a company doesn't have three basic components:

* Quality Products — a CRM system can't get customers to buy products they don't want.
* Brand Name — a CRM system can't rehabilitate a disreputable brand name or a brand that's fallen on hard times.
* Capable Channels — channels that can't function with success individually usually won't function successfully when brought under the umbrella of a multichannel model. Adding a CRM system to the mix is like writing good code over bad code — it still won't work.

✪ WHAT'S THE BEST STRATEGY?

A business can dip its toes in the water by developing independent CRM capabilities within various parts of the company. These are called point solutions. This approach can be less jarring to the company. Many times such CRM projects are based on function-specific needs. For instance:

* Marketing may clamor for a CRM system to plan, execute, and monitor marketing campaigns and perform database marketing.
* Sales may take the CRM path to help it with lead management and to provide sales force automation capabilities to support field sales.
* Logistics and fulfillment may add CRM to their bag of tricks to support mass customization and to provide up-to-the-minute information on product in transit to the customer.
* Call center and technical support may lean on CRM tools to help in the

deployment of sophisticated telephony and information systems to provide ongoing customer service, and up-selling and cross-selling capabilities.

The advantage of taking the separate capabilities route (i.e. point solution or modular approach) is to provide a means to support specific CRM projects, which fit within an overall CRM strategy. This approach allows the corporate culture to gradually shift from a product-focus to a customer-focus business model. A point solution can still allow a company to obtain some or all of its stated CRM objectives. For example, a sales and marketing point solution could enable these departments to focus on customer retention and on increasing the share of the company's overall customer base instead of acquisition and market share. Or perhaps customer service automation tools could help to identify and take advantage of cross-sell and up-sell opportunities, thus helping to achieve a customer enhancement objective.

But taking this route without proper planning for the long term can be costly. This type of approach to CRM doesn't provide for valuable customer data flowing freely across departmental boundaries; but rather this method means the data is still kept in individual departmental silos.

To achieve the Holy Grail of CRM — real-time 360-degree view of the customer corporate-wide — requires a massive effort. But it's only through such effort that integration across all departmental boundaries (marketing, sales, logistics and fulfillment, and customer service and support and so forth) can CRM heaven be achieved. Where everyone with a customer-centric interface can have access to the latest information on the customer — profile, behavior, and expressed needs.

With the right CRM system, a company can differentiate the service they provide with every customer contact. Case in point are loyalty programs where the "gold" customer expects and receives superior service (remember, 20% of customers represent 80% of a company's profits). Thus, the business community strives to treat their best customers as special, giving them intimate and personalized service to ensure that they do retain these highly valued assets. That's not to say that customers who are down a rung or two won't receive good service (some might be among next year's gold or platinum group), but rather they may receive differentiated service as resources permit.

☻ COST CONSIDERATIONS

A successful corporate-wide CRM initiative can be expensive, very expensive with multimillion dollar tabs. But, before giving up on the idea of a CRM initiative because of the cost, note that a 2001 Data Warehousing Institute survey of more than 1600 business and IT professionals found that almost 50% had CRM projects in the works with budgets of less than $500,000. This indicates that CRM doesn't necessarily need to be a budget-buster.

For some businesses even this level of investment is out of the question, but

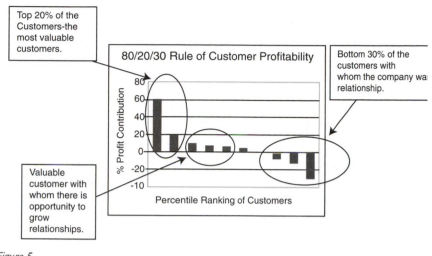

Top 20% of the Customers-the most valuable customers.

80/20/30 Rule of Customer Profitability

Bottom 30% of the customers with whom the company wa relationship.

Valuable customer with whom there is opportunity to grow relationships.

% Profit Contribution

Percentile Ranking of Customers

Figure 5.

don't give up. If a company wants to play in the same ballpark as its well-heeled competitors, there is a solution — rent a CRM solution through a HSP (hosted solutions provider). A HSP can also allow a business to test a CRM solution over a period of time, permitting it to measure anticipated ROI before taking the plunge.

Or perhaps a company could initiate a less technical-intensive CRM project. Say a personalization project — by targeting a specific group of customers with specialized messages. Take a period of time to evaluate the results and if they are positive, go forward with a more aggressive CRM solution. However, if the experiments don't work out, the company hasn't dropped a bundle of money and the CRM champion still has his or her job.

Clearly target the CRM objectives and heed the results of the needs analysis. If a business fails to stop and take stock of its assets — human and technical — it can find itself overpaying for a CRM solution that is either way beyond its needs or doesn't match its needs.

Need More?

Before touting the need for a CRM initiative, it would be nice to know what kind of statistics can be used to convince the boss that it is a worthwhile investment. (Or, if you are the boss this section can be used to convince your conscience that it is a worthwhile solution.) Pay heed!

Glenn Berkwitt, VP of corporate marketing for Primus reports that one of its customers increased its profits by 30% after installing one of Primus' CRM solutions. Another Primus customer boasted that it was able to increase its "first call close rate" by over 30% after the first call compared to using its former legacy system.

Quick & Reilly reported some noticeable benefits after the first year of operation of its CRM initiative. According to a company spokesperson, its CRM system helped to produce, at minimum, a 15% increase in new accounts.

A survey conducted in mid-2001 by WebSurveyor and CRM vendor, YOUcentric, shows that when it comes to ROI, CRM is on safe ground. Forty percent of executives participating in the survey reported that their company's CRM solution has yielded solid ROI, with another 40% adding that they have seen some ROI. Only 20% claimed not to have seen any ROI at all.

According to quarterly surveys of Siebel client companies during 2001, on average they have seen a 12% annual increase in revenue per employee, a 20% rise in worker productivity and a 20% boost in customer satisfaction.

Other studies have shown that companies that increase their customer satisfaction rate by only 10% realize up to a 10% increase in sales.

✪ IN SUMMARY

Remember, the purpose of the evaluation process is to understand if the organizational structure, decision-making process, training programs, and funding can support the business requirements of a first-class CRM initiative. Such an evaluation process helps prevent the company from sabotaging the CRM initiative before it's even launched.

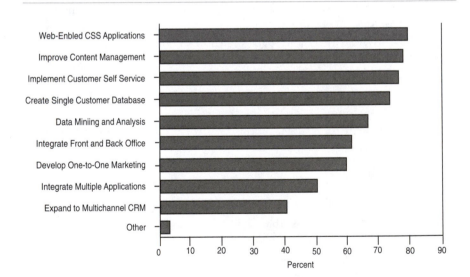

Figure 6. - Key CRM Initiatives. Source: Gartner Dataquest (March 2001).

Chapter 3
Preparing a Business Case

Before giving the green light to a CRM project, executive boards require a compelling business case to support the economic benefits of a CRM system. Any major CRM investment must have specific financial goals that can be achieved relative to the current state.

CRM projects are typically complex since they affect major areas within a corporation. CRM cuts across all functions of a business — from the marketing division through to sales and customer service, dispatch and more.

A successful CRM project depends upon a strong foundation and a good strategy. These are delivered *only* through the right design and technology, embodied by a well thought-out and structured implementation plan or blueprint. The strategy must be supported by a thorough business case that can satisfy all who must support the project. This calls for a systematic, disciplined and thoughtful analysis of the need for a CRM system within a company.

How to make a compelling business case for CRM? By blending the benefits of a CRM system with a realistic projection of cost, contingencies, and schedules. Include detailed risk and ROI analysis, all targeting the company's unique qualities and goals.

Show the big picture — how a CRM system will add value to the business. Bring the department managers onboard by showing them what they will gain from a CRM initiative. Remember that political and cultural issues are the "hot potatoes." So create benefits for all — sell early and sell often. It's the winning formula for getting influential managers and executives on the side of CRM.

Begin the business case by describing CRM and how it can help the company best manage its different types of customer relationships. Then follow with a demonstration of why the system could be a strategic winner. Quantify the benefits (in detail) and provide specific information about how the system will achieve those benefits. Conclude the business case with supporting data for ROI and outline the risks associated with the project.

Note: Use the business case as a working document. This will allow everyone to gain increased confidence in the strategy, statistics and their impact as there becomes a greater understanding of the potential of the CRM initiative.

Bear in mind that during the construction of the business case that most executive boards insist on using quantitative analysis to guide their decision-making.

I present the difficulties of building a business case first so the readers realize what they are up against. The most significant challenge faced when building a business case for a typical CRM initiative is a realistic assessment of the benefits (based on estimates of potential revenue improvement), which are usually very difficult to quantify since CRM benefits don't start accruing until late in the investment cycle.

The Business Case Outline

What follows provides suggestions and recommendations on how to construct a successful business case.

A) A business case in support of a CRM initiative provides three facets of information:

  ✳ Presents clear and factual information to support the decision process for the initiative.

  ✳ Identifies and establishes strategies to manage risk associated with the initiative.

  ✳ Identifies the factors and progress measures impacting the success of the initiative.

B) A comprehensive business case must show that the CRM evangelist has provided for:

  ✳ Realistic estimates of investment, execution schedule, and potential contingencies.

  ✳ Project (and program, if need) leadership that has experience managing complex projects.

  ✳ Committed executive sponsorship ready to solve the inevitable leadership issues that typically occur with any CRM initiative.

Step-by-Step

A successful business case builds support for the resources and changes needed for the CRM initiative to succeed.

1. Start by providing a clear working definition of CRM and its flavors, including the point solutions proposed for implementation.

2. Before identifying areas where savings can be made, estimate the change between the current status quo and what the future status quo will be. A

systematic analysis of the current state will provide a baseline from which to measure improvements.

3. The next step is to calculate the benefits to demonstrate how a CRM initiative will support the company's business strategies. Put hard numbers against those benefits.

4. Verify the potential impact of the CRM initiative. When it comes to showing the potential benefits of CRM within the corporate walls, industry benchmark information can be of value. Another method would be to take current operating costs and present them in a manner similar to benchmark information.

5. Follow with a calculation of the costs of the initiative through a detailed budget that covers pre-implementation costs; integration, implementation and deployment costs; and post-implement costs (support and maintenance belong in this category). Include both hard (definite, quantifiable and proven) and soft (indefinite, hard to quantify and/or to prove) costs.

The most challenging task will be the development of a realistic estimate of the implementation costs and the potential benefits of the CRM initiative. It's easy to overlook significant transition costs that can eventually jeopardize full implementation of a CRM initiative.

Don't forget to include identification of the key cost targets and methods for monitoring progress against goals and objectives. This could include a sensitivity analysis detailing the impact of the CRM initiative's objectives by exceeding or missing project targets.

6. Assess the impact of risk, but also show how risk can be minimized.

✪ THE CONSTRUCTION PROCESS

Discuss the type of CRM initiative being presented. For instance, the business case might be for a pilot project, the first of many CRM projects that eventually will be integrated to form a corporate-wide CRM initiative. It could be a project to feed real-time information into call centers without bringing other departments and divisions into the fold. Or perhaps, the CRM initiative consists of a specific field sales project that needs little input from the marketing department. Or it could encompass a one-to-one relationship management system that requires a complete overhaul of the entire corporate structure.

Regardless of size or scale, how invasive or isolated the CRM system might be within the company, demonstrate why the CRM initiative is a strategic winner by quantifying benefits at a detailed level and specifying how a CRM initiative will achieve those benefits.

Explain what CRM is and what it does. Give the audience a close look at some of the many flavors of CRM and summarize vendors, consultants, and systems integrators necessary to each project.

Next, calculate the benefits by explaining how a CRM initiative can support the company's market and customer strategies. As illustration: The major benefits that the CRM initiative provides are a 45% decrease in costs through the ability to provide a quick response to competition and/or changes in the market place. Be specific.

Then put hard numbers against the benefits. This includes calculating project costs and presenting a budget for the project including both hard and soft costs.

Examples of the typical hard cost (not comprehensive) are:

* Hardware.
* Communications infrastructure.
* Network/data transfer.
* Licensing fees.
* Vendor costs.
* Consultancy costs.
* Systems Integrator costs.
* Roll out service.
* Training.

Soft costs might include:

* Configuration/Installation (in-house).
* Strategy/process (typically in-house).
* Project management (in-house or outsourced).
* Data processing (moving, recoding and cleansing data).
* Other in-house costs (sales and marketing staff, management input, etc.).

Note: For every $1 that a business spends on CRM software, it also spends between $2 and $5 on implementation and maintenance. This may sound costly, but the investment could be well worth it.

Costs

So, what will the CRM initiative cost? The pre-implementation (evaluation, presentation of a business case, mapping of a strategy), vendor selection, the implementation and deployment and then the post-implementation costs varies greatly from company to company and is dependent upon the type of CRM solution(s) being introduced. A point solution or pilot project that doesn't require integration of sundry departmental systems will usually cost less than a project that requires more integration effort. (It should be emphasized that a point solution is implemented within the framework of a corporate-wide CRM strategy.) The most costly will be a full-scale CRM initiative.

Return on Investment

How to present the return on investment (ROI) of a CRM system? How to assign specific value to the many benefits which are soft or intangible and hard to quantify? How to deal with the numerous other independent variables? Although not easy, it can be done. When presented in the right vein, the ROI can demonstrate and build support for the changes needed for a CRM initiative to succeed.

Most ROI calculations are multipurpose — they satisfy the need to convince management and to obtain funding. For example, the questions that most senior managers will ask are, "Will the system add value?" and "What's in it for me?"

Canny CRM enthusiasts will prove the CRM initiative can add value, i.e. that it's worthwhile and will increase profits and shareholder value, and that the investment is better than other alternatives. This is the point in the business case to show how CRM will support the company's business strategies and why this specific CRM initiative is the best option.

Present the "real" costs and benefits of the CRM initiative. For example, a 15% increase in direct and indirect revenue generation through (say) the addition of a more loyal and profitable customer base, expansion of the wallet-share of existing customers and/or improvement in cross-selling and up-selling initiatives. Or perhaps a 20% improvement in the productivity of marketing efforts and a 5% increase in cost efficiencies of service channels and a 10% increase in effectiveness of sales channels.

Reveal the indirect revenue impact of a CRM initiative. This could include increase in brand and customer equity, customer retention and satisfaction and increase in the overall percentage of profitable customers. Maybe, also show indirect cost reduction brought about by a CRM initiative, such as increased speed to market for new products/services; or the reduced fulfillment and customer response errors through increased synergy between the company's systems.

Explain how the CRM initiative can maximize scarce resources or assets — not just cash — such as customers (afterall they are the company's greatest asset), staff (the company's most expensive asset), time, market share or even economies of scale.

Another method is to present financial metrics established prior to implementation so as to determine feasibility/justification of the investment. These metrics should track and determine the economic value contribution after the CRM initiative is up and running. Use the metrics to quantify the amount and timing of the economic return relative to the investment required for each CRM project. (If there is no baseline data prior to installation of a CRM system, before and after comparisons will be impossible.)

Next, address the implementation itself — its milestones and deliverables. While this data should be relatively easy to come by — it's difficult to use with most CRM initiatives. The problem is that a *major* CRM initiative isn't *just* an implementation project — it's a capital investment in a strategy to change the way a company does business. If the business case relies solely on such internal measures, the perception is that success is defined by completing a project on time and on budget — not by the results generated from the investment.

Measure the Benefits

The real critical success factors and performance indicators for CRM are the eventual improvement in sales and profitability. But these are based on traditional quantitative metrics, which are influenced by many other independent variables that rely on long-term goals. To present a more definitive idea of how well a CRM project or initiative is working out requires the measurement of the causal relationship of any increase or decrease of sales and profits in the short-term to the implementation of a specific CRM solution.

One way to demonstrate hard revenue improvements is to use, as an example, the prospect of sales increasing without the benefit of CRM through, perhaps, more direct mail, telephone work, people, or advertising. The benefit derived from the proposed CRM system would be the costs saved by negating the necessity to increase sales through one of more of those methods.

Other examples of traditional quantitative metrics might be:

* Reduction of costs through customer self-service solutions.
* Benefits from channeling certain sales to call centers and/or a website.
* Cost savings achieved through time efficiencies.
* Reduction in customer churn.
* Improved efficiencies in the value chain through linkage of customers to suppliers.
* Improved focus on high value customers.
* Improved understanding of the individual customer purchase life cycle.

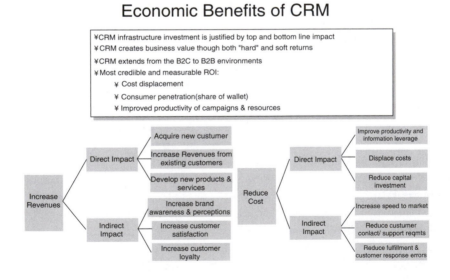

Economic Benefits of CRM

¥CRM infrastructure investment is justified by top and bottom line impact
¥CRM creates business value though both "hard" and soft returns
¥CRM extends from the B2C to B2B environments
¥Most crediible and measurable ROI:
   ¥ Cost displacement
   ¥ Consumer penetration(share of wallet)
   ¥ Improved productivity of campaigns & resources

Figure 7. CRM affects revenue and cost structures both directly and indirectly.

* Initiation of improvement in cross-sell and up-sell systems and processes.
* Improved understanding of customer profitability.

Of course, no matter how such figures are arrived at, everyone needs to acknowledge that there are just too many other independent variables (price/performance of product/service, competitive market conditions, productivity of staff, fulfillment and delivery performance, management focus and commitment, etc.), which can influence traditional financial performance measures.

For example, a gain or loss in customer satisfaction and loyalty can be measured through customer surveys, repeat sales, call to resolution data, and returns. But when trying to quantify customer care quality (whether the measure used is satisfaction, retention, frequency of purchase and/or average transaction volume), many other internal and external variables can still come into play.

Nevertheless, it's advisable to list the benefits customers will derive from a CRM system, such as:

* Improved response time to customer requests.
* Improvement in percentage of delivered product that meets customer requirements (resulting in reduced cost to the company for product returns).
* Immediate access to order status (resulting in reduced cost within the company's call centers).
* Greater breadth of solution options.
* More responsive technical support.

The truth is that many of the benefits of a well-oiled CRM system are both tangible and intangible. As a matter of fact, the main benefits of CRM are soft and intangible, meaning they are difficult to quantify. The hard, financial paybacks of CRM are realized over varying periods of time and some can't be measured for decades (such as the benefits of a customer-focused culture).

Discuss the Risk

Then proceed to the impact of risk. Discuss such risks as:

* Lack of core competencies within each CRM project user group (i.e. marketing, sales and customer service staff with less than stellar computer skills).
* The chance that there will be a lack of discipline and commonality across the company's sectors (e.g., there may be unforeseen difficulties in integrating all departments and divisions).
* Cultural and political differences throughout the company (for example, sales not happy with the call center servicing a portion of the customer base).
* Speed (if a project takes too long to launch it may stall due to outdated process, loss of interest, etc.).
* Management turn-over can cause delays and/or create friction within a company and produce havoc within a CRM project.

Need Some Incentive?

When looking at the various statistical data on CRM and its impact on the business community, companies that provide superior service to their customers enjoy certain advantages over their competitors that have failed to adopt a first-class CRM initiative. For example, these companies:

* Grow twice as fast as their competition.
* Experience a 6% annual market share growth vs. a 1% share loss (i.e. they take customers away from their competition).
* Charge 10% more for their products and still manage to take customers away from their competition.
* Enjoy a 12% return on sales versus the 1% average return inflicted upon their competition.

Use a favorite search engine and read industry publications for timely, specific statistical data.

✪ IN SUMMARY

Although building a business case can be problematic, it can and must be done. It must, however, be done well. A poorly formed business case will not only preclude the funds necessary for the CRM initiative; perhaps more important, management might approve the budget while *failing to appreciate the necessary* changes that must be made (cultural, process and business practice) for the CRM initiative to succeed.

A compelling business case must demonstrate why the proffered CRM initiative is a strategic winner. It must give the big picture in such a way that senior management grasps the implications of CRM and, thus, provides the necessary support for the changes that a CRM strategy wroughts throughout the corporate structure.

Chapter 4
Creating the CRM Strategy

I t's fairly easy to put together a committed CRM team, and with a good business case, it's not too difficult to get the executive suite's buy-in for a CRM initiative; yet the actual creation of a CRM strategy can be quite elusive. Nevertheless, the benefits of a well-developed CRM strategy — increased sales, new ways to differentiate the company and its product in the marketplace, the ability

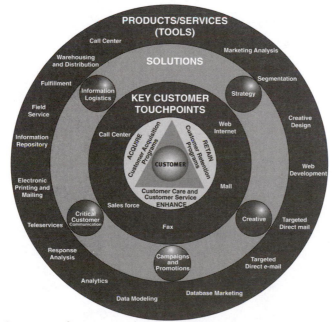

Figure 8. Components of a company-wide CRM strategy.

to absorb new business methods (i.e. the Internet) — add substance to the cost savings used to support most CRM initiatives.

Customer relationship management itself is a business philosophy that adopts a strategy. Both place the customer at the heart of the corporate culture thereby compelling a re-engineering of a business's processes and activities. It is NOT an application, a technology or a suite of products.

A CRM strategy usually requires a cultural shift — aligning the company, its employees and systems towards its customers (a customer-centric philosophy) and away from its tried-and-true product- or process-centric philosophies that focus on cutting costs and improving efficiency through optimization of internal processes and automation of back-office functions.

A customer-centric philosophy allows a customer to interact with a particular company when, where and how the customer wants, and allows the company to capitalize on *every* interaction. With a customer-centric business model in place, a company learns how to keep its best customers. (The "best customers" represent the core of any business and should be the target of carefully orchestrated customer retention efforts.) A CRM strategy also ensures maximum efficiency and effectiveness without driving costs to excessive levels.

A customer-centric philosophy recognizes that customers are more than the end of the value chain, they are part of a product/service delivery continuum. It acknowledges that all customers are NOT created equal and that development of a deep knowledge about the *right* customers is critical. A customer-centric busi-

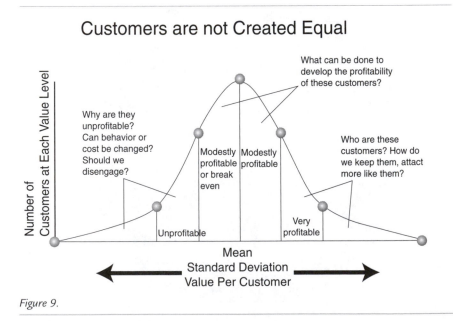

Customers are not Created Equal

Figure 9.

ness model allows a company to understand the unique value it brings to its customers, individually and as a whole.

The right CRM strategy allows a company to combine its information systems, policies, process and employees so it can attract and retain profitable customers. While businesses will continue to acquire new customers, they know its essential that they focus on growing the *right* customers. By managing the business's relationship with the right customers, a company can increase its profitability.

CRM software and products are the tools used to implement this strategy. But, they must be woven into the very fabric of the company's business strategy, not bolted on to it.

⊙ WHERE TO BEGIN

CRM is all encompassing. Although a bottom-up strategy can optimize isolated departmental needs and provide point solutions, it cannot change the corporate culture. That must come from the top — the executive suite. No one department, call center, or IT division can drive the cross-functional process changes required for a company-wide CRM initiative.

To adopt a customer-centric corporate culture, the business community must also adapt its technological architecture so as to realign and even reinvent its business and work processes prior to jumping on the CRM bandwagon. But, while technology is essential to the implementation of a CRM strategy (it underpins the processes), it is only part of any CRM project. The adoption of a customer-centric philosophy and its supporting strategy, along with change management techniques that educate of the end-users, are even more critical to a successful CRM initiative.

A successful, useful, and profitable CRM initiative always starts with a business strategy (versus a CRM strategy) that can serve to drive change within the company. This internal change drives the move away from a product-centric philosophy toward a customer-centric philosophy. While a product-centric business model can reap significant benefits from CRM technology, adoption of a customer-centric business strategy is necessary to reach the pinnacle of that CRM technology. Yet, even a business that continues to sport a product-centric business strategy must establish quantifiable objectives for a CRM initiative because that's still one of the keys to an effective CRM strategy. The other is high-level executive backing.

Indeed, the same holds true for a company that starts its march toward CRM with a point solution (such as sales force automation), within the framework of a customer-centric business strategy. There is still a need for a definable objective and high-level executive backing (for the funding, if nothing else) for the project to achieve success.

Note: A recent study revealed that the CEO was directly involved in successful CRM initiatives more than 40% of the time.

The implementation of an effective CRM capability usually requires an upgrade in the business's database technology that stores, updates and maintains the customer information. In most instances, there is also a need for data cleansing prior to launching the CRM system. Therefore, the capital expenditure can be high, even when a business takes only a tentative step toward introducing CRM into their business mix.

CRM tools offer new options and opportunities. But make sure that the CRM initiative is channeled into the right project or projects. This is one of the most formidable and important tasks in business today.

Find a Strategy that Fits

Each company must find a CRM strategy that addresses not only its specific business philosophy and strategy, but also the company's current needs and priorities. Let's look at business philosophy and strategy first. The CRM team needs to determine whether the company's philosophy and supporting strategy is based on:

1. *Product-based selling.* This is where a company gathers information about its customers' transactions and performs simple analysis of the variables. (Note: This strategy doesn't offer any real insight into a specific customer's relationship.)
2. *Managed service and support.* The company's focus is on applying customer service to the sales process to improve customer relationships, such as setting up a call center and/or help desk.
3. *Customer based marketing.* Here focus is on the customer. This method allows a company to make different offers to different customers and identify individual cross-sell and up-sell opportunities, maximizing customer profitability.
4. *Personalized relationship marketing.* This requires that the business collects and analyzes extensive information about its customers. It then uses this knowledge to offer personalized service, i.e. one-to-one marketing. (To support this strategy, the business must have an IT system that can support a corporate memory of each customer — how and whenever they enter the corporate environment and details of each interaction.)

Customers are the force of CRM, so take time to evaluate the company's customer strategy. First, understand how the company's customers define value. Find out why the existing customer base buys from the company. Does the company offer:

* A unique product or service?
* A price that's better than anywhere else?
* Faster delivery of its product or service?
* A relationship that is so strong that the customer has no desire to go elsewhere?
* Combination of any of the above?

Now consider the company's current needs and priorities and identify the specific situations that cry out for a CRM solution. Such as:

Inconsistent Marketing Initiatives. If a company has multiple sectors marketing to the same customer base, it has a problem. This typically occurs within a corporate environment with a product-centric philosophy where customers receive various offers from different product lines with potentially conflicting messages. Customers often think they must decide between one offer or another versus the initial marketing objective of enticing the company's customers to opt for all products being marketed through these efforts.

Multiple Channels. Many companies find that their customers increasingly use multiple channels in their daily contacts. Although these customers still use the traditional channels (mail, retail stores, outside sales, faxes, telephone) they also use new channels (Internet, kiosks, wireless, email) The business community must adapt their processes to accommodate the channels its customers utilize.

Data. Many businesses have multiple databases containing varying degrees of information about their customers. Multiple databases have the same issues as multiple marketing initiatives — no one-view of the customer, which limits the creativity behind retention and loyalty programs. Also, although a company's data warehouse may have a solid architecture, quality data, and work quite well, many times it's not used. With CRM in the picture, the data warehouse is once again in the forefront.

Globalization. Due to the Internet, globalization has snowballed into an avalanche of competitive efforts. Thus, some within the business community have begun to have trouble acquiring new customers and their customer attrition has increased. They find that the old ways — mass marketing, product-centric marketing etc. no longer generate the required results. They must, instead, find a way to encourage many of their customers to *proactively* generate more revenue.

A CRM strategy allows a company to understand where customer information is located, how it's going to be used and how it might flow through an integrated CRM environment (from the source to a data storehouse and to the customer through any channel the customer chooses to use). Top business priorities must be agreed upon by all involved. The CRM strategy should be used as a roadmap to show how the company will get from "here to there" by combining strategy, technology (features, functionality), personnel, timelines, milestones, and costs to make a CRM initiative a reality.

⊛ UNDERSTANDING THE TECHNOLOGY

The appropriate CRM technology is the enabler. Thus, the CRM strategy and its attendant technology should fit like a glove. But, the CRM team needs to understand the CRM technology, it's not a piece of software that can be snapped into an IT ecosystem. Instead, CRM technology through *integration within a company's*

IT infrastructure enables a business to develop, archive, and share customer information throughout a business to, for example:

* Identify its customers' specific needs.
* Offer its customers a personalized view of the business.
* Deliver efficient and standardized customer care.
* Identify the business's most profitable customers.
* Identify the customers who are most at risk.

CRM technology helps businesses manage and organize customer touch points. With the right tools, properly customized and integrated for a specific IT ecosystem, a business can respond to a customer's needs instantaneously (e.g. keeping its catalog, web and sales team continuously updated on product and pricing changes). Customized product offerings and quotes on the fly are a piece of cake. For example, a company can set up CRM tools to send existing clients reminders about upcoming service requirements on their equipment. Or it can implement true one-to-one marketing, sorting through online customer profiles and purchase histories to adapt new offerings to each customer's individual preferences.

Customer relationship management applications can include, but are not limited to:

* Call Center Automation.
* Campaign Management.
* Contact Management.
* Data Warehousing.
* Email Management.
* Field Service Automation.
* Knowledge Management.

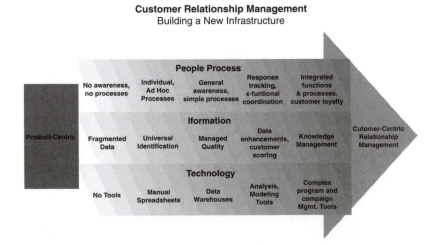

Customer Relationship Management
Building a New Infrastructure

Figure 10.

✳ Marketing Automation.

✳ Personalization.

✳ Sales Force Automation.

Technology is rapidly advancing. For instance, new communications technology can connect road warriors with the company's IT systems; the Internet can deepen self-service options and personalization efforts; and advances in telephone systems make virtual call center operations possible. Technology allows a company to make the most of each customer contact. Customers expect to speak immediately to a person, who already knows all about them. They expect a website or telephone system to provide complete information that meets their needs.

But it doesn't stop there, while technology that exploits the information in front-office applications (sales, marketing) has been around for quite some time, technology that can capitalize on information held in back-office systems (accounting, purchasing material management, distribution) is relatively new to the market.

That's just the tip of the iceberg. CRM tools also enable customers to self-select assistance through a company's website from anywhere, at anytime, day or night, seven days a week. Customers can gather information about products, update billing information (even pay bills), review account balances, and check orders on their own — and then immediately receive more personalized help from a call center agent, if needed, with a single click of a button. This capability allows businesses to improve customer service while reducing costs and improving productivity. This is all CRM.

Managing Data

The CRM team should consider how complete the business's current customer data is in relation to the degree to which that data must be used to offer its customers individualized or personalized service.

Does the company need CRM tools to help it to better understand the customer data it already has by converting that data to provide better information about its customers' behavior and interests? There are products that can help to identify what additional information is needed and then determine how this knowledge can be used to guide future customer interactions. There are applications that can generate quality leads for sales teams, identify and target high-profit customers, and reduce redundant work and data reentry errors by sharing a common data source throughout the company.

Let's look at the appropriate technology as it relates to data flow. An inevitable consequence of most CRM strategies is the need to collect more data and then derive additional information from that data. If a company shifts to a customer-centric strategy, it must plan for more data and greater integration of that data — from both its front-office (e.g. call centers and other customer-facing applications) and back-office (accounting, order process, logistics and fulfillment, for example).

Once that is accomplished the system must be able to distribute this parsed data to more people (employees and customers). The most common technology options are:

1. *Tactical Databases* These are used to support specific needs, such as a marketing campaign management or sales force automation. Most companies also carry out analysis of product sales and other transactional data directly upon operational system databases. In this case, the scope of the analysis is limited because data isn't linked with data within other operational systems.

2. *Data Marts.* Most data marts cover specific subjects and are kept separate from the operational systems; i.e. built solely to store all the data that's collected to support, for instance, a sales and marketing CRM initiative. Moving from a product-centric to a customer-centric philosophy usually requires more advanced data mart components.

3. *Data Warehouses.* This type of system serves as a single repository of data, as opposed to multiple data storehouses. Once the data warehouse is up and running with clean data, it provides a single version of a customer or group of customers, company-wide. The appropriate query and analysis tools and data mining software enable a business to better understand its customers, allowing it to plan more advanced CRM strategies.

4. *Integrated CRM Solutions.* This is a system that has the ability to incorporate both data warehousing and web-based technology. It usually requires the integration of several customer-facing systems and e-commerce applications, with data warehouse technology. This enables the company to provides its customers with personalized and coordinated service across all channels. Integrated CRM solutions can tightly link both front- and back-office applications.

Thus allowing a company to gather data that:

* Provides responses to sales and marketing campaigns.
* Concerns shipping and fulfillment dates.
* Provides customer sales and purchase data including notations on increase/decrease in volume and time of year most purchases are made.
* Compiles account information including payment/credit history.
* Integrates web registration data.
* Provides service and support records.
* Takes into account demographic and web sales data.
* Calculates why a customer buys from the company.
* Determines if a customer is a advocate for the company, refers others to the company and even whether a customer is influential within his or her specific industry or niche.

An Actionable CRM Strategy

Upon examination, some CRM failures can be blamed on the lack of an actionable CRM strategy (i.e. a well-defined series of steps or processes for implementing CRM, along with measurable CRM goals and objectives). An actionable CRM strategy might read as follows:

* Increase new customer acquisition through web-based channels by X% over the next N months.
* Reduce customer turnover rates from X% to Y% over the next N months.
* Achieve cross- and up-selling targets of $X over the next N months from customer service and technical support departments.
* Improve customer satisfaction measures by X% and reduce order placement costs by $Y over the next N months.

The reason I underscore the importance of an actionable CRM strategy is because it's the best way to garner top-level executive sponsorship, which is the only way a company can make the necessary changes within the corporate environment.

Developing an actionable CRM strategy will significantly improve the chance of a successful CRM project by:

* Identifying the CRM drivers from a high-level strategic perspective.
* Mapping CRM drivers to specific functionality areas within the company.
* Assessing the functionality with the most impact on the company's business objectives.

Such a strategy begins with defining CRM from the business's standpoint. Specifically, what are the benefits from different phases of a customer relationship? For instance, if a call center is the focus of a CRM project, the CRM team needs to ask a few pertinent questions, such as: How does the company benefit from the customer relationships garnered through its call center? How does the CRM project influence these benefits? In other words, how will CRM in the call center bring additional strategy, features, functions and analysis capabilities that will enhance the benefits already enjoyed?

Features and Functionality

It's CRM technology that provides the necessary features and functionality to achieve the objectives of the CRM strategy. (At a fairly high level, the functions and features offered by CRM technology address one or more of the customer relationship phases.) CRM technology spans a broad range of *functionality*. These can range from management systems, such as contact management, campaign management, dealer/distribution management, pipeline management and field service management to CRM *features* incorporating product configuration, telemarketing, customer interaction centers, customer analysis (real-time and historical), self service and personalization. Thus, the CRM field is large and continually expanding as new kinds of functionality and features evolve to serve the customer.

Viewing the features and functionality available in products offered by the CRM vendor community is a daunting task. The list can cover nearly every functional area within a business. Cull the list, look at the three distinct phases that every business goes through when dealing with its customers:

* *Acquisition* — understanding what, when and where the customer's needs lie. This phase is usually found in sales, marketing and even the call center.
* *Retention* — meeting the customer's needs and expectations. This phase usually includes services provided by customer service departments, accounts receivable and R&D, but also might include order entry and dispatch.
* *Enhancement* — cross-selling and up-selling products and services to existing customers. Interaction at any customer touch point can potentially sway a customer's decision to purchase additional services and/or products.

When looking at CRM technology from this perspective it isn't quite as intimidating.

Customer Acquisition

Every business seeks to acquire "profitable" customers. Sales force automation and campaign management technologies are examples of tools that support customer acquisition. As most of the readers know, on average it costs a company ten times as much to acquire a new customer as it does to retain a current customer.

However, few companies can gain insight into why a prospective customer may opt to *not* do business with them. Companies that go the extra mile and develop appropriate applications to capture the information required to understand the needs of its prospective and existing customer base will gain a competitive edge.

A CRM initiative to increase customer acquisition doesn't have to be a million dollar project — a website with customer self service and product information can help a company acquire new customers and retain existing ones. Add the ability to configure products and the advantages become more distinct. Recent research conducted at the University of Texas found that close to 70% of the companies surveyed reported an increase in revenue by providing customer access to IT systems for configuring orders or researching information online.

Customer Retention

Once a company acquires a group of customers, it can retain that group by making them feel special through customer recognition. Even something as mundane as when my UPS delivery person always addresses me by name and makes some individualized comment when delivering a package can have a positive affect on a customer relationship. More elaborate strategies, such as product offerings, loyalty programs, service innovations and quality improvements are now possible because there is technology that allows a business to "listen" to its customers. One

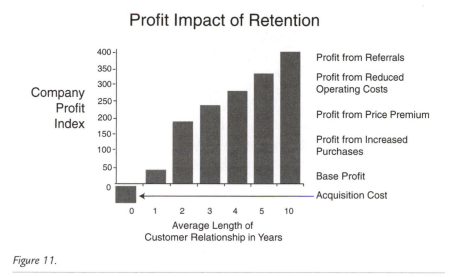

Profit Impact of Retention

Figure 11.

example of a CRM-related perk is when a broker gives its customer a special telephone number to call where that customer can bypass the normal hold time. Another is creating a sense of community for users of a product or service, such as what Prime Radio Products, Compaq Computer, and Kodak provide.

Constant customer turnover or churn is a major factor driving the quest for customer retention functionality. Loyalty programs, such as Barnes & Nobles' Reader's Advantage program, the airline industry's frequent flyer programs, and American Express' numerous awards programs, are examples of customer retention tools.

Admittedly, most companies are fairly successful at customer acquisition and retention. If they weren't they would be out of business. However, most could do better. That's where CRM really shines. Accepted industry figures show that even if a CRM initiative only reduces customer churn by 5%, the company can significantly increase its profits.

Customer Enhancement

Finding ways to grow a company's customer base through CRM-based acquisition and retention techniques are all well and good, but executives and managers want the ability to do more. At the top of their wish list is the ability to easily cross- and up-sell related products and services to the company's existing customer base without annoying the customer. Granting that wish can be difficult. Despite having terabytes of customer data, most businesses can't extract meaningful information from their order entry database to (say) allow their call center agents to implement a cross/up-sell strategy. That's where CRM-based enhancement tools come into play, transforming all customer touch points into profit centers. Companies are just beginning to realize that all of their customer touch points have the potential to be profit centers. Imagine turning traditional back-office cost centers — customer service, help desks, technical service, billing and shipping — that normally have post sale interactions with customers, into profit centers. Just implementing the technology that would allow this one CRM feature can have a positive affect on a business's bottom line.

Post sales performance and web-based service play a vital role in the enhancement phase. In the current corporate climate, businesses do not have the luxury of adding personnel to meet their customers' increasing demands for more service and support — enter CRM-based tools that lean heavily on web-based delivery methods.

The company, like many businesses, probably has endless transactional information about its customers distributed throughout multiple databases islands, such as its contact management system, order entry system, accounts receivable, warranty and the list goes on and on. In the past, this customer data couldn't be converted into information that could help enhance the company's relationship (selling products and services that add value to an existing relationship) with its customers. Now various CRM-related technologies enable these islands of corporate data to communicate with one another. This is the first step in providing customer relationship management, which, in turn, allows a company to enhance its relationship with its customers.

The 21st Century business community is placing the customer at the center of its universe, whereas the 20th Century community was more preoccupied with achieving operational excellence in its production and service delivery processes. The right CRM technology can provide the features and functionality that allows a company to make the customer the focal point company-wide.

◉ THE ROAD TO SUCCESS

First and foremost, a successful CRM initiative requires commitment from the top down. Without it, even a company that has heavily invested in CRM may stare failure in the face.

Chief Causes of CRM Failure:

A CRM initiative can fail if a company finds itself in any of these situations:

* Lack of executive sponsorship.
* Poorly defined strategies, goals and objectives.
* Too much change required within the corporate structure without executive buy-in for the change.
* Failure to take into consideration the impact of the project upon existing business process.
* Ineffective project and resource management.

* Lack of awareness and training.
* Technology inadequacies, which can include:
  * Unconnected processes.
  * Mismatched and/or incompatible technology within the infrastructure.
  * Confusion over what tools to use.
  * Dirty data.
* Lack of leadership to reel in recalcitrant employees.
* Financial constraints due to lack of commitment in the executive suite.
* Poor fiduciary control.

Second, a workable CRM strategy. While developing a CRM strategy can be difficult, when CRM is viewed incrementally by going through the customer relationship phases of acquisition, retention, and enhancement, it's easier to envision an actionable CRM strategy, giving the CRM initiative a better chance for success.

Third, a CRM strategy should be based on the channels that *the company's customers use*. These channels are identified through research. The web and Internet technologies may or may not be vital components. For instance, if the company provides pharmaceutical supplies to the medical community, inside and outside sales personnel are its main customer channels. In this instance, a website would be used almost entirely as an information conduit, not as a sales channel. However, if the company sells software, a website might be its main sales channel, backed up by retail stores and call centers.

Note: When developing a CRM strategy, don't limit the search to specific technologies or vendor, because the solution needed may require a range of technologies, which no single vendor can accommodate.

Actionable Planning

Mapping out specific customer relationship objectives in a succinct fashion is only half the battle. The CRM team must supplement such customer relationship phase with associated implementation timelines. Together, these factors can dramatically improve the chance for a successful CRM project.

Conversely, the typical CRM team is charged with implementing strategies such as "give the company a 360-degree view of the customer." This needs to be translated into an actionable plan, which measures areas of various customer relationship phases that are going to be improved with this 360-degree view.

The next step is to prioritize which of the measurable areas are most important to the business. Without a prioritized list, the chances of a successful CRM implementation are drastically reduced.

Let's take this a step further and walk through a couple of typical scenarios.

After designing a CRM strategy with actionable steps, a transformer manufacturer comes up with specific objectives in the customer retention and sales enhancement areas. The business line managers reveal that they are losing customers to a new company that has undercut their pricing. The product manager feels that sales of the company's new line of transformers would increase exponentially if existing customers (who've hit community forums with tales of their difficulties with an older transformer model) were made aware of the stellar performance record of the company's new transformer. One of the least costly and most effective ways to do this would necessitate training the technical support staff who are dealing with the woes of the older model, so they could make the sale.

Of course, this now brings another set of problems. While the company wants its technical support department to cross-sell, it must also deal with how this will affect the sales department, which faces the prospect of losing sales commissions. Also, it must determine the risk of the technical support department abandoning its primary focus of technical support and becoming sales commission-oriented. Then there is the question of whether the technical support manager will be evaluated by the number of support calls successfully closed or by sales targets. These issues have to be resolved before the technology can deliver on the business objectives.

Thus, when processes and incentive structures are modified to support enhancement objectives, even a technologically inferior application can deliver value. By contrast, even the best CRM technology will fail, if the processes and incentives are not changed to support the company's customer relationship objectives.

Articulating a clearly defined customer relationship objective and having management work through the feasibility of a CRM initiative, allows the process and incentive changes needed to support this objective become clear for every affected department. Being specific also has the benefit of allowing the IT department to manage scope creep and focus on delivering features and functionality to achieve the stated objectives.

Similarly, a web-based business may need to decrease the number of abandoned shopping carts and improve its customer retention rates. To achieve these objectives requires real-time customer analysis and personalization functionality, which requires a different IT ecosystem to support the necessary CRM technology. Or perhaps a company, such as a PC manufacturer, may find that product configuration feature is needed for customer acquisition and retention, which requires, among other things, a change in business and workflow processes. To achieve these objectives require that the same scope of thought and planning, as related in the previous scenario, occur prior to implementation of CRM technology.

As these examples show, it's important to examine the company's strategy and objectives and how it maps to specific CRM functions before making any acquisition or implementation decisions. The advantage of mapping specific objectives in this manner is that it lets everyone, including technical and non-technical personnel, understand the impact of the proposed CRM initiative. This sets the stage for selecting CRM technologies based on the business drivers as opposed to implementing CRM technologies without a clear definition of the business objectives or strategy. Now a CRM project has a good chance of success.

Delivering Customer-Centric Functionality

Developing an actionable strategy, combined with regular reality checks, should help to "short list" the necessary CRM functionality for those decision-makers addressing a company's immediate business objectives. Executives are more likely to understand and support organizational change issues during the technology implementation phase — when they are in pursuit of clearly articulated objectives.

Recall that previously I mentioned the vast array of CRM functionality — from contact and lead management to call center and field service functions. If, at the end of developing and mapping an actionable strategy and conducting reality checks, the CRM team cannot whittle the list down to a few functions for the initial CRM initiative, then clearly the chances of the company enjoying an ultimately successful corporate-wide CRM implementation are remote. The amount of organization and process change needed to implement CRM technology increases dramatically as the number of desired functions increase. Even a large, well-defined CRM initiative needs to be broken down into smaller, manageable projects.

✪ AVOIDING POTENTIAL LAND MINES

Mapping and selecting a CRM strategy from the customer relationship phases of acquisition, retention, and enhancement has the benefit of a simple and clear articulation of goals and objectives that the entire company can understand and rally around.

Many CRM initiatives (even the multi-million dollars projects) have quietly stalled or failed as corporate management searches for business benefits and as employees shy away from technology they say doesn't help them in their day-to-day duties.

Some companies, humbled by past brushes with failures and/or stories of CRM struggles, have opted to take a more measured approach. While vendors may encourage large CRM initiatives, starting out small with a pilot project actually may be the best CRM strategy. Mapping out several small projects as part of a well-structured CRM strategy can provide very successful CRM results.

Many CRM enthusiasts are caught in a confusing triangle. They are pressured on one side by CEOs desperate for a quick CRM fix, on another by vendors falsely promising a CRM nirvana and on yet another by problematic end users who are often slow to adopt the precepts and technology of CRM. Still, everyone must keep their head. Don't let your company be among the group that has been so bewitched by a vendor's slick marketing strategy that it jumps into a CRM project without clear strategies, objectives, needs analysis and support from top management.

It's vital that the CRM advocate finds a way to cut through vendor hype and ensures that the company's employees can and will use the new CRM applications. The best way to defuse these potential land mines is to partner with top executives and management. They can successfully sell the new CRM systems internally and help institute and promote company-wide training programs so that the employees feel comfortable with the new system.

As the reader has probably noted, CRM initiatives have many ways they can go wrong. Paralysis is the number one pitfall. Another is creating the entire physical data model based on every source system and every field and then trying to map all of this to a data warehouse in one fell swoop, rather than incrementally. Other problems that can cause a CRM initiative to stagger might include:

* ✱ Never-ending technology evaluations and RFPs.
* ✱ Lack of understanding of how the project can move from the strategy stage to the implementation stage.
* ✱ No clear understanding of who is in charge of funding and where the funding is coming from.
* ✱ No one person in charge of verification that the ROI is accurate.
* ✱ Failure to enlist the aid of an experienced project (and program) manager.
* ✱ A lack of a strong CRM evangelist.

✪ KEY STRATEGY ISSUES

How can a company develop and implement a CRM strategy? What will be the role of senior executives in implementing a successful CRM initiative? Where can the company find the personnel with the right skills to implement a successful CRM initiative?

For most businesses, CRM is a new direction — a new strategy that leads to greater profitability through the creation of customer loyalty and a stable customer base. Today's business community realizes that if they don't take care of their customers, the chances are that someone else will. Yet, many think that CRM is all about technology, usually because it's pushed by software vendors who find that talking technology is easier than discussing culture and business process changes that must be implemented before any CRM solution can be considered a success. Never forget that while CRM isn't possible without technology, a CRM initiative within a product- or process-centric environment does little to enhance a business's competitive quotient. Such CRM projects are severely handicapped because they are instituted to support a business's current culture and processes, which by their very nature aren't customer-centric.

Then there's the inability to identify what needs to be done and lack of attention to creation of the right collaborative culture. This brings about an uncoordinated project. The lack of strategy and direction results in confusion all around.

In this type of environment, corporate politics and self-interest run rife, CRM systems are built that aren't capable of supporting the supposed objective of the project, and in the end there is no positive change in customer satisfaction or loyalty. Alas, money and time has been wasted.

My advice: Although the goal of a CRM project is to move the company toward a customer-centric philosophy, it should be considered an end user project, not an IT project. And, bite off something manageable, then manage the scope and create business value.

⊛ IN SUMMARY

It's not easy for any business to keep all of the promises it makes to its customers during the course of day-to-day business, and even more difficult is keeping these promises while containing costs. The right CRM strategy can ameliorate this situation by tying together a business's existing systems, personnel, products and services. Any company, no matter the size, can begin to implement CRM features and functionality into its business model by providing a well-structured CRM strategy that implements effective CRM one step at a time — starting with areas where the business can realize immediate benefit.

With the right CRM strategy, a company can profit from a good understanding of where its customer information can be found, how this information will be used and how it might flow through an integrated CRM system (from source to data storehouse out to customer touch points). It should also have a clear idea of what technology is in use and how its diverse systems relate to one another.

Finally, change management consistency must be a part of any CRM strategy. The company must decide on the amount of change at the outset of the CRM initiative and then take the appropriate action. The CRM initiative should have a strategy that's properly mapped, an adequate budget, realistic timelines and milestones, and a plan for training all staff affected by each CRM project.

No two businesses are the same. New techniques to facilitate change may need to be developed from a variety of disciplines. Use this chapter as a guide, not as a cut and dried blueprint.

The fight for customer loyalty is on, and if a company's CRM strategy is well thought out, it can boost a company's stature and revenue.

Chapter 5

A Program of Projects

The best way to surmount the challenges of the 21st century economy is to develop a comprehensive program of CRM — project-by-project. This offers the best opportunity to design and build the necessary CRM capabilities, while at the same time keeping an open framework wherein new technology and business strategies can be accommodated in the future.

CRM initiatives are disruptive. They call for a strategic change within the corporate environment, which means working in a different way. Few businesses, except perhaps new startups, have the right infrastructure to manage relationships with their customers without massive reorganization and rebuilding.

Note: A CRM project is defined as an undertaking with a definite starting point and a defined objective(s). Once the objective(s) is reached, the project is complete.

For a CRM strategy to succeed, changes are required in the personnel structure, company culture, customer and internal processes, reward and recognition systems, employees' skills and competencies, information management, measurement systems, and technology.

Very few CRM evangelists present a business case that convinces the executive board to give them the *authority and responsibility* to identify ALL that needs to be done for a successful company-wide CRM initiative. Consequently, many within the business community have failed to achieve the promised benefits from their CRM initiatives.

To overcome a down turn in customer satisfaction, loyalty quotient, retention rate, etc., a company must look to management techniques that can coordinate all the work involved in building the necessary customer management capabilities into one integrated CRM initiative. This is true whether the CRM initiative is implemented in one fell-swoop or in smaller increments through point solutions, a modular approach or pilot projects.

Note: *Change management is most frequently stated as the greatest challenge of a CRM initiative. A survey conducted in mid-2001 by Knowledge Systems & Research, Inc. (a marketing research and consulting firm) confirmed this. That survey found that change management was the most challenging aspect of implementing a CRM strategy. Second was technology integration (it trailed almost 9 points behind change management), and the third was lack of IT resources and skillsets.*

If a company only needs to improve the effectiveness of its sales force by introducing key account management or improve its call center's effectiveness; project management will fit the bill. However, these types of CRM projects, in and of themselves, do not move a company from a product-centric culture to the necessary customer-centric culture. That's not to say that by using quick wins through individual, discrete CRM projects, a company can't obtain valid short-term CRM objectives. It's possible. But, to implement a customer-centric business strategy, these individual projects must exist within a larger framework — a program of projects — otherwise the initial quick win might gradually fizzle out, depending on the size of the company and the objectives of its CRM strategy.

Does the CRM Initiative Need Program Management?

Any CRM initiative needs, at a minimum, a project manager. *Only* companies, which have a CRM initiative on the drawing board that fits at least three of the following criteria, should consider implementing its CRM initiative as a program.

✳ The company's current business strategy does not place an emphasis on CRM.
✳ The business case is difficult to prove, but upper management believes that changes are needed to stay abreast in today's competitive market space.
✳ For the CRM initiative to be successful, deliverables must be established, prioritized and agreed to by many diverse departments within the corporation.
✳ The benefits of the CRM initiative need to be managed as the work evolves.
✳ The changes necessary in moving to a customer-centric environment will be disruptive to the current corporate culture.
✳ For the CRM initiative to be a success, current processes will need to undergo many changes, and new processes must be implemented.
✳ The company, as a whole, does not understand CRM, or how it will affect them.
✳ There is likely to be opposition to the change and, therefore, strong leadership will be required.
✳ An e-business model is necessary to support the CRM initiative.

Note: Not every CRM initiative needs to be ran as program.

The decision on how the CRM initiative is managed will far outweigh the consequences of how well the CRM tools perform. Also, the tighter an initiative's implementation schedule, the greater the need for experienced project management.

While a CRM initiative takes place in a corporate environment, the approach to corporate management and project and/or program management are very different. Consequently, a credible case must be made for including professional project and/or program managers in the CRM budget. It may be useful to point out that nine out of ten successful CRM initiatives are professionally managed from start to finish.

Many corporate executives' initial response to the question of adding a project or program manager to the CRM team will be, "we've done quite well without these managers in the past, so why bring one onboard now?" The CRM advocate's response should be, "to provide professional project and/or program management capabilities to the implementation process that emphasize communication, project planning, team coordination, status reporting and issue resolution." But that's not all, it should be pointed out that perhaps the most significant differences is philosophical. Corporate management philosophy is to make do with as few employees as possible. Understaffing is a death sentence for most CRM initiatives. A CRM initiative is risky and therefore needs a margin of surplus (to take care of contingencies that will bubble up — if for no other reason).

While both the project management and program management approaches are used to deliver changes that can improve customer satisfaction and conversely, business performance, these two management disciplines are used in different situations, have different focuses, and require different types of leadership. The real value in using one or both of these management disciplines is in the quality of the end results, and the avoidance of unnecessary delays and costs. It's difficult, therefore, to place a value on such management approaches, at least in traditional accounting terms. However, as one of Golub's Laws of Computerdom points out, "A carelessly planned project takes three times longer to complete than expected; a carefully planned project takes only twice as long." Ergo, a well-run project could result in huge cost savings.

A "given" is that CRM initiatives and even individual projects will almost always take longer than expected. Some of the things that can wreck havoc with timelines are:

* Items that haven't been considered and there is always something that's overlooked.
* The tendency for the CRM advocate to be overly optimistic when presenting the project to the executive board for approval, which can create the perception that a CRM initiative takes too long to reach completion.

* Risks that were not fully appreciated or even disregarded during the planning stage, can and do rear their ugly heads.
* Failure to start the project on time.

Another "given" is that CRM initiatives will always cost more than expected. Again, optimism during the approval stage is the main culprit. While overly optimistic cost estimates that result in under-estimating the actual cost of a project or initiative is politically motivated to ensure approval, it can cause problems down the line. Then there are schedule delays — each schedule delay will translate into added costs. Another culprit is planning deficiencies that may not show up until the implementation process begins, which can substantially add to the costs of a project.

◎ PROJECT MANAGEMENT

Project management is used when a specific outcome (such as enhancing customer management facilities in a call center environment) is required in a set time frame. Projects typically have clear and easily defined benefits and ROI, so management focus is on minimizing risk and cost. Projects are also generally managed within boundaries (e.g. building a call center or implementing a sales force management system) and don't spill over into other areas. Thus, they have an easily identified and defined scope.

The goal of project management is to achieve success in as efficient manner as possible. CRM projects require coordination of resources and unless this is carefully planned, the company may end up with less than optimal results.

Note: Project management can also be a subset of program management. One such example is when a company initiates a pilot CRM project to test the waters prior to investing in a company-wide system. Another is when a company draws up a defined company-wide CRM strategy that calls for instituting the over-all CRM initiative over time by implementing numerous individual department-centric projects using point solutions or individual modules from a suite of products.

The Project Manager

A project manager is given the authority and responsibility to undertake the work necessary to implement a specific CRM initiative to obtain a specific CRM objective(s) without constant surveillance and supervisory intervention.

A project manager should:

* Understand the project's goals and objectives.
* Know why the project is being undertaken.
* Understand why the project is necessary at the current point in time.
* Grasp the risks involved with the project.
* Present a realistic budget and delivery timeline.
* Understand the tangible and intangible benefits of the project.

* Have an opinion as to how successful the project will be as far as the company is concerned.
* Know who his or her supervisor is and from whom directions will be received as the project progresses.
* Build a project team through the use of in-house staff, if possible. However, if the available in-house staff doesn't possess the required skills, then outside resources must be found and brought onboard.
* Develop a project plan.
* Identify any and all information and resources required for the project to succeed.

A project manager needs a clear idea of what constitutes a successful conclusion of the project, and works toward that end. The project manager has the added responsibility of helping the program manager and the CRM team to understand the details of the project when changes are necessary that require additional planning, scheduling, communication (technical and public relations), reports, coordination and supervision. In such a way, a project manager can take responsibility for leading the project towards its stated and agreed upon goals and objectives.

Finally, a project manager's most important function is that of communication. The project manager must be able to clearly articulate the status, needs, and direction of the project to the program manager, CRM team, and the technical and support staff.

Note: *A project manager works in coordination with a program manager, when the project is part of a company-wide CRM initiative.*

The Project Plan

The project plan is a key document and the responsibility of the project manager. It's used as the basis of reference throughout the term of the project. An effective project plan:

* Sets out the goals, objectives and scope of the CRM project (its deliverables).
* States how the goals and objectives will be obtained.
* Provides the standards or quality control to be used during the integration and testing process.
* Provides the design outlines (schema) and components (CRM tools, other software, middleware and integration needs, including hardware).
* Assigns team responsibilities and provides coordination of the work. This allows the team members to understand their participation and responsibility and commits them to the success of the project.
* Tracks and steers progress of the project through appropriate initiatives.
* Outlines a budget for the project that establishes realistic cost estimates for every component of the project before it's committed to implementation.

* States the resources required for implementation (people, material, technical, space, existing assets, external interfacing, etc.).
* Provides economic and cash flow projections.
* Outlines areas of uncertainty (risk) and contingency plans. By their very nature, CRM projects can lead to both risks and opportunities, so risk management is important. Risks must be identified, analyzed and then responded to in order to reduce adverse time and costs implications. At the very least a project manager should prepare contingency plans as a precaution.

⊙ PROGRAM MANAGEMENT

Program management supports strategic change, which allows a business to link vastly different initiatives (e.g., corporate cultural changes including personnel indoctrination, changes in a business's infrastructure and management, and rebuilding of the business's infrastructure).

Program management is the best way for a company to control the portfolio of projects that come with a company-wide CRM initiative. But that is just the beginning. Program management allows a business to prioritize its CRM projects to ensure the company obtains maximum benefit from its CRM initiative while securing efficient, coordinated implementation. Without program management it is difficult to achieve corporate benefits and manage personnel in the new corporate environment that CRM brings to the fore.

As illustration: Instituting a marketing management program is a *project*. Repurposing a company from a product-centric environment to a customer-centric environment is a *program*.

Overall, program management involves identifying, prioritizing and linking initiatives, many of which may become projects within the program, but not all. In this case, the management focus is on efficiently attaining corporate benefits (customer satisfaction and increased profits) through a new strategy (morphing the corporate emphasis from a product-centric business strategy to a customer-centric strategy).

Program management can provide the right environment for change to happen, particularly in terms of the employees' attitude and behavior. The nature of a program means longer time frames, such as what would be required for a company-wide one-to-one CRM initiative. Non-CRM examples of the necessity of program management are Bell Atlantic's massive rebranding to Verizon and perhaps, a cable company rebuilding its network to offer IP-based broadband services.

The Program Manager Skillset

A company should not assume that one of its project managers can assume the role of a program manager and run a long-term CRM program without first obtaining additional training and experience. Although program management techniques do draw on project management abilities, there is also the need to effectively coordinate strategic change; this isn't found among the skillset within the project management genre. Program managers must have the confidence, experience and influence to lead a business through the confusion that massive change can bring about.

Project management skills place great emphasize on detailed planning and leadership that can motivate a team to achieve defined goals. Whereas *program management* skills rely heavily on organizational leadership and communication skills that can build and lead cross-functional teams, play in the game of corporate politics, requiring diplomacy and the ability to obtain executive and upper management support, and the kind of top-notch communications skills that are needed for public relations.

The program manager works closely with individual project managers in all stages, including planning. However, program management only provides the framework, the detail is delivered by the project managers. *Program managers* usher in the benefits of the new strategy. *Project managers* manage cost and reduce risk to ensure ROI is achieved through their own project and then this spreads across the multiple projects required in a program management environment.

The Difference

Project management is like the process needed to get a plane from one specific airport to another at a specific time and date. The project succeeds if the plane lands at the designated airport on schedule.

Program management requires the skills to get a host of different vehicles and drivers to arrive simultaneously at one place at a specific time and date. An example would be a person in charge of getting Madonna and her entourage from a distant location to a concert hall at a set time and date. This could require scheduling and managing the operators of planes, trains, automobiles and trucks to ensure the right people and equipment are at the right place at the right time.

If the company envisions a corporate-wide CRM initiative, then each CRM project should be ran within a CRM program format. Now, CRM program managers are a rare breed and the company will probably need to call on outside help — not to run the program, but to provide training and skills transfer. The outside advisor or consultant isn't brought in to run the program, but rather to work with a strong project manager and to teach that person program management skills through teaming, knowledge transfer and on-the-job training.

The Program Manager's Responsibilities

The responsibilities of a program manager can vary widely, depending on the company, the program and its projects, and the program manager. For example, he or she might be put in charge of stewardship of the CRM initiative and one of the leaders of the CRM team. This could include: compliance with the company's overall objective; appointment of project managers; approval of the scope of all projects and their overall success indicators; ensuring that policies and procedures are in place and followed; and finally, project risk assignment, which might consist of the full management and the financial consideration of:

* Political motivations within the company.
* Obtaining continued financial support for each and every project.
* Changes in the company's cultural and physical environment
* Any top-down notice of urgency.
* Potential interference in obtaining team members, contracts, etc.
* Backup strategies for high-risk areas affected by any CRM initiative.

A program manager might also responsible for implementation of:

* The stated new corporate customer-centric strategy.
* The overall procurement strategy for the CRM initiative.
* Effective public relations to gain support for the CRM initiative.
* Efficient record keeping for the overall CRM program including comprehensive documentation and archiving chores.
* A plan to ensure the orderly transfer of the care, custody and control of each CRM project upon completion, such as employee training.

The Program Management Plan

A program management plan can differ from company to company; but in general, a program management plan's objective is to coordinate the various ongoing and planned projects that make up the CRM initiative. It can be used to obtain approvals, such as executive board approval of the details and funding to proceed to subsequent phases (projects) of the program or to obtain funding and approval for necessary add-ons for an existing project.

A program management plan is essential to instituting an effective company-wide CRM initiative. Such a plan should ensure that:

* All project objectives within the CRM initiative are carefully assessed, enunciated and prioritized.
* A logical and systematic approach is used to obtain the objectives of the CRM initiative.
* There is a realistic cost/benefit analysis for the overall CRM initiative.
* The CRM initiative has a clearly articulated overall strategy.
* The project managers have presented realistic timelines, schedules and budgets.
* Definition of the overall CRM program scope is set out in separate phases or projects prior to commitment to the overall CRM initiative.

✪ CORPORATE OVERSIGHT AND CONTROL

The CRM initiative should have executive oversight. This executive or executives should be either a member of the executive board, a director or a senior manager who uses information technology on a daily basis. While CRM initiatives require a dedicated project (and possibly a program) manager, a control executive is also essential — for sign-off at enumerated executive control points, if nothing else. For instance, at major milestones within a project's timeline, the project or program manager would present to the executive certain predetermined deliverables, thus giving the executive the final say in determining whether the CRM initiative receives a "go" to proceed to the next milestone.

It costs around ten times as much to implement a change within a CRM initiative after integration work has begun as compared to recognizing the need for such change during the planning stage. Hence, the control executive is charged with exercising a high level of control over the shape and timing of each CRM project within a CRM initiative. In this way, the executive is in charge of ensuring that the overall CRM initiative develops in a manner consistent with the company's outlined objectives.

Note: The control executive should have the authority to order modification of a CRM project if the company's objectives change.

The control executive also has the duty and opportunity to provide morale boosts through his or her enthusiasm for the CRM initiative and confidence in the CRM project and its project team. During the formal approval sessions, the control executive can reiterate strong commitment to the CRM initiative and confidence in the project and/or program managers' ability to perform their jobs successfully.

In addition to the control executive, there may be a need for an executive oversight board that is responsible for items such as:

* Ensuring that policies and procedures are in place for overall program management including substantive changes within the CRM initiative.
* Each project is prioritized according to necessity or return on investment, justified by a business case, and aligned with the company's overall strategic business plan.
* Releasing major project funding on an approved project plan basis.
* Appointing a project control executive.
* Ensuring that all CRM projects and the overall CRM initiative are completed.
* Promptly conducting any necessary project post reviews.
* Halting, re-planning or scrapping projects that have lost their way.

✪ IN SUMMARY

Regardless of the CRM initiative being implemented, project and/or program managers are valuable assets. They give a company the best chance of ensuring its

CRM initiative progresses according to plan. To maximize the return on investment of any CRM initiative, the initial implementation should be brought in as close to budget as is humanly possible while at the same time adhering to all project timelines and milestones. Implementing CRM through project and/or program management is a safeguard — it's as close as a company can get to a guarantee that its CRM initiative will be a success.

Chapter 6
The People Factor

The objective of a CRM initiative is to provide a business with flexible, powerful and appropriate options for managing the various aspects of its customer interactions. To be successful CRM must also improve the end user's (employee's) experience. CRM cannot escape the people factor — to succeed in winning customers, a company must first "win" its employees. When planning the CRM strategy, the employees' reaction to change, their learning curve and their commitment to adoption of a new system should be brought into the picture.

Ignoring the human element can bring a CRM initiative to its knees. Customer relationships are not managed, controlled, or even enabled by CRM tools, instead they're managed, controlled and enabled by people — the company's employees. In the same vein, most employees feel they already *understand* what the customer wants, so they don't want an added burden of keying information into some piece of CRM software to tell them what to do and how to do it. Conversely, personnel who *don't know or care* what the customer wants, might use the CRM system, but only to perpetuate old workflow and old habits. While neither situation constitutes an outright CRM failure, either will cause a CRM initiative's projected ROI to take a nose-dive.

CRM technology provides customer-related details. Yet, in the end, the fate of the customer relationship is in the hands of the employee. The technology is put in place to support the employees so they can work with the company to improve customer relationships. If employees refuse to use it or if it doesn't support what they really need to do — the company ends up with some very expensive, seldom used, screen icons. To quote Lee Iococca: "Start with good people, lay out the rules, communicate with your employees, motivate them and reward them. If you do all those things effectively, you can't miss."

CRM can't deliver on its promise if management can't master the internal relationship puzzle. The very nature of CRM demands interdependence within the corporate structure. It requires everyone within the company to think and act in new and non-linear ways.

No individual or department can remain an island — more helpful relationships means more information shared — helpful relationships cross departmental boundaries. When the corporate environment demands teaming, problem-solving across job boundaries, knowledge-discovery skills, and cross-departmental networking, it gives birth to new relationships that share knowledge and information.

✪ MANAGING THE CHANGES

As new relationship skills are learned, there is need for diligent oversight. Many times such skills are overlooked in the rush to focus on the more tangible aspects of CRM (such as new technology or exciting marketing and sales campaigns). CRM thrives in companies that learn to nurture helpful internal relationships. With this cultural change, information can be shared faster and problems can be resolved quicker, motivating employees to buy into the CRM vision.

Changing the way a corporation works disrupts its people, which, in turn, brings on resistance (active and/or passive). It's critical that the members of a CRM team find ways to minimize such impediments — it slows a project down and can place the CRM initiative at risk. But, changing peoples' behavior can be difficult. During the upheaval that CRM brings to the corporate culture, the staff is asked to think and behave in different ways and even may be given new roles. Therefore, along with implementation of a CRM solution:

* ✳ Train sales staff to become more relationship focused.
* ✳ Encourage service staff to be more proactive.
* ✳ Product managers need to think about solutions and work together.
* ✳ Marketing staff needs to think more about campaign results and analysis.
* ✳ Recognize that the IT organization and its staff, through their input and performance, provide a competitive advantage.
* ✳ Company-wide, the personnel need to learn to focus relentlessly on the customer

The CRM team must exhibit a little sensitivity. The change to corporate culture and staff behavior can imply to some that what was done in the past was wrong — their training wasted. Their skills no longer valued — this can be demoralizing. Then in some companies, the employees' experience is that they are catapulted out of their "comfort zone" into an alien world where they have to struggle to speak the language, deal with incomprehensible currency, and understand strange customs. In such a situation, the employee will want to fall back into their more familiar work pattern and environment at every opportunity.

Management Buy-in

No CRM strategy can go anywhere without bringing department managers onboard the CRM bandwagon at the outset. They must motivate their staff to *want* to deliver more value to the customer through CRM. For many corporations this means staff behavior and attitude must change, which requires finding ways to overcome the reluctance to adopt a new, automated CRM system and all that it necessitates.

Importance of User Cultural

Gartner rates "user culture" (employee attitude) as one of the leading causes of CRM failure. If on a scale of one to five a CRM team finds that the company's user culture rates below a four, then postpone the process and study the situation a bit more.

Rating	User Culture
5	The employees are demanding access to the new system.
4	The employees are motivated but unsure of how to use the new system.
3	The employees feel they must use the new system "or else".
2	The employees only show "some interest" in the system's capabilities.
1	The employees are completely unmotivated to use the new system.

The results of the rating exercise may help the company to decide that staffing adjustments are required. Or, after the rating exercise is completed, the CRM team just might find out that the company isn't ready for CRM — at least at this point in time.

In the call center environment, it's typical to find CRM systems that go unused because the staff can't or won't use them. The cause may be because the management's ethos of "get them off the line quickly to reduce costs" has led to poor customer service practices and reduced staff morale, or it may be the result of poor training procedures or even employee resistance. Whatever the case, the situation must be remedied, if the CRM initiative is to score a "win."

Another typical problem area is the sales department — sales people are an independent lot — they don't adopt or accept change easily. If a company's CRM project is a new sales force automation (SFA) system that provides the latest technology, enabling it to acquire new customers and boost its bottom line, but it can't get the sales force to use the system, it's money down the drain. That is, unless that sales force can be convinced to buy into the CRM project. When the end user is comfortable with a CRM system (such as a SFA system), the long-term success of the system is greatly enhanced.

The average sales person will evaluate a SFA system based on criteria such as:

✴ Is there a steep learning curve?

✴ Is the system easy to use?

✴ Will the new system save time and reduce overhead?

✴ Will it simplify customer/prospect contact and fulfillment?

✴ Does it enhance communication with the customers/prospects?

✴ Will it increase productivity?

If there are more than one or two negative responses, it will be difficult to win over this group of employees — the CRM project can be blocked or even derailed. To avoid such mishaps, make an all out effort to ensure that the CEO and top sales and marketing executives (or top executives of the reluctant department) are actively involved in the CRM initiative from the "get-go". Full cooperation from management is necessary if a company is to obtain a satisfactory "buy in" from the affected staff. Management must be the driving force in defining and prioritizing the primary business issues and in providing effective communication to all of their staff.

⊙ COMMUNICATE

Focus on the end user from initiation to completion of the CRM project. During the planning and implementation process, the CRM team can be too focused on getting the job done, giving too little attention on how the job will be done. This can leave the individual employee feeling uninformed and uncertain about what is happening and how it will affect him or her. This results in a "bunker mentality" where the employees worry that they will be expected to do more with less. So they stonewall.

People resist what they don't understand. Get the end users involved from the start. Involving the greatest number of end users possible, early in the process, broadens the knowledge and appeal of a new CRM project and creates a sense of ownership in the project. Let them test different solutions, then listen to and address their comments. (The average employee knows what the problems are, has a strong opinion about the current situation, and want to improve their working environment.) Such an approach makes the employee a part of the solution, rather than just another problem.

⊙ TRAINING, AND MORE TRAINING

Perhaps the most important aspect in assuring the CRM initiative is successful is addressing how the end users interact with the system. Because CRM systems are built in large part by IT resources, many times the end user is one of the last items considered in a CRM strategy plan. Don't let time run out. Take the steps prior to the implementation process to provide optimal training and support resources for the affected employees.

Even a best-of-breed CRM system may be rendered useless if the staff isn't properly trained. Training can't be a one-time affair; it must be an ongoing process. If the company doesn't train its employees properly in the use of the system and continually reinforce the training effort, then the investment in strategy, software, hardware and development are wasted.

To help overcome employee resistance, implement a training and support center that can provide not only intensive training on the system, but also system demonstrations prior to implementation, and "hand holding" during the rollout phase (this might last up to six months).

○ A "CHANGE MANAGEMENT" PLAN

People matter, so understanding change management and its use in a CRM initiative is important. CRM often provokes large, disruptive cultural change within the business environment. It involves a shift in focus from the products (internal) to customer requirements (external) and it's happening at the same time as seismic changes in the markets are moving power from the business community to the customer. That is a lot of change to accept, let alone to understand.

The best way to accomplish such jarring changes is to put in place a *change management plan*. Change management is the use of systematic methods to ensure that the organizational change necessitated by a CRM initiative is guided in a cost-effective and efficient manner.

Whether the company's employees see change as a positive or a negative — resistance is inevitable. However, with careful planning and consideration of the human element of change, a CRM initiative can avoid some of the pitfalls that plague many CRM projects. Pay heed to a GartnerGroup research report that found the "people factor" is always among the top causes of CRM failure. Contributing factors include:

* Hands-off leadership, where managers know nothing about what the customers want.
* Rewards, recognition and incentives based on efficiency rather than effectiveness.
* A culture that has a relentless focus on internal matters rather than the customer.
* A lack of skills in understanding customers and supporting a customer infrastructure.

Companies need to put resources into and budget for helping their staff to understand these changes and then to enable them to contribute to a new, more profitable way of working. With a change management plan in place, it's possible to minimize employee resistance by showing them what they can gain and encouraging them to willingly modify their work practices.

CRM change management is not user training at the end of a technology implementation nor a "change manager" putting out communiqués on implementation progress. Change management starts when the idea of CRM is first raised with the executive board and key influencers. It should be based on a CRM team that can:

* Communicate the objectives of the CRM initiative in a clear, concise manner.
* Establish a good relationship between the end users, IT staff, and top executives.
* Approve rewards (financial and/or non-financial) to encourage use of the new CRM system.
* Ensure there is the necessary support (technical, training and help desk) for the system.
* Put into place a method that allows end users to comment on the effectiveness (or non-effectiveness) of the new system and suggest improvements.

One of the ways to accomplish these tasks is to use a phased plan that moves the company from a customer-aware, but fragmented culture to a more team-orientated and customer-intimate culture, and finally to a customer-collaborative culture. At each stage, relevant techniques are used. Be patient, replacing old work habits with new beneficial patterns take time.

Spreading the Message

The first phase is spreading the message. The message of change should issue from the executive suite, since they can bring "carrot-and-stick" techniques to bear. Make the aim of the first phase of a change management plan to be soliciting agreement from key executives and influencers on what CRM really means to the company and the company's commitment to the need for change.

In this phase, a "political map" that shows key executives, key influencers, CRM advocates and CRM detractors is invaluable. People who are heavily connected to the initiative politically should be addressed in different ways. For instance, key executives and influencers are good at spreading the message. Advocates and detractors are both potential candidates for the CRM team. (Always keep detractors close — as they are won over they can be used to demonstrate the benefits of CRM.)

Change Impact and Adoption Strategy

The next phase is to establish the best approach for corporate-wide change. Create a change impact and adoption strategy. Previous staff satisfaction surveys are a good place to start. If one doesn't exist — take a survey. This strategy, which is part of the change management plan, should address specifically and in detail:

* Who will be affected by the change?
* To what degree will they be affected?
* What is their likely reaction to the change?

Note that customer-facing sectors are more likely to be affected than back-office areas, but no one should be left out. Not surprisingly, members of security, reception, delivery and porter staffs can offer invaluable insight on a company's customers, since they actually have more interaction with customers than is often realized.

Employee Buy-in

Now it's time for the employee buy-in phase. Give employees a stake in the change — this is an opportunity to allow them to contribute to improvements and at the same time for the company to show them that the CRM team values their opinions. Start the buy-in process by asking for employee input. Take a CRM survey that not only gathers the crucial information needed in developing a comprehensive CRM system, but also incorporates the end user's prospective. The survey should ask questions, such as:

* What functions do you perform?
* What kind of data do you use on a daily, weekly, monthly and irregular basis?
* How do you interact with the company's customers?
* What information should be made more readily available to you to enable you to better understand and help the company's customers?
* What are your reporting responsibilities, needs and requirements?
* Are you involved in obtaining leads, lead tracking, lead follow-up, data transfer or other daily details? If so, how can they be improved to increase your accuracy and efficiency?
* How can your communication with customers be improved?
* How can your administrative, reporting and scheduling requirements be reduced?
* Are you involved in customer outreach activities, such as telemarketing, direct mail or direct email campaigns? If so, how can they be improved and/or your efficiency be improved?

In this way the CRM team can determine the tools each end user feels they need for success in developing a meaningful relationship with the company's customers. These are the people who perform jobs on a daily basis on behalf of the customer. Their input is worth listening to — they know the problems and have ideas about what changes are needed to improve the company's relationship with its customers.

Collaboration

Once the CRM team has garnered employee support, the next phase is collaboration. This phase can create a real understanding of what CRM means to the corporation and make the staff feel they are contributing to the change. The tools of success during this phase are team building, workshops for idea development, customer feedback to guide the work, and a model demonstration of CRM principles. Since the average person only retains about 60% of a visual message and 25% of an audial message after a 48-hour period, an ongoing internal public relations *strategy* can increase the odds that the employees will understand the CRM initiative, and accept and use the new system.

Success is attained when there is a "sandwich" effect, with the entire corporate structure being repurposed — from the top down and the bottom up — at the same time. This phase can quickly move the company and its staff through some of the known change cycles that typically cause problems.

Acceleration

The final phase is acceleration. In this phase the momentum of change can be increased. Techniques should be employed to speed up the spread of knowledge about the benefits of change, what customers are asking for and how the staff will benefit. Now is the time for new ways of working and thinking to be built into various programs, including communication channels and methods, rewards and incentives, training, and coaching. (Coaching is often better than training as people learn best when new ideas are related to their experiences and are immediately useful.) Since the very essence of CRM is that information sharing needs to be fast, focused and integrated, the end users' skillset needs to reflect the technology they are using.

This is where teamwork is encouraged. Intranet technology, virtual team rooms, knowledge-sharing communities and employee portals are some of the tools that can be used during this phase.

⊛ IN SUMMARY

When the heat is on, there is the tendency to neglect good employee relations practices in the rush to get the job done. This attitude will cause the CRM initiative's ROI to plummet due to low productivity, low morale, high turnover and lost opportunity, resulting in more time being spent on internal management and not on managing customer relationships.

Realize from the outset that there will always be a learning curve before everything can come together to bring about a new functional way of doing business. There will be a need for tough people decisions — there is always a portion of any employee group (management and non-management) that can't or won't accept new ways of doing things. Yet, if a company can't manage its employees how can it expect to manage its customers.

Once the change is well in hand, the emphasis can turn from an internal focus to an external one — the company's customers — to build a culture of "relentless customer focus."

Chapter 7

Increasing Customer Loyalty

The *quality* of market share (customer loyalty) deserves as much attention as the *quantity* of market share (product sold). This means shifting the emphasis from selling one million widgets, to that of ensuring that the company's widgets are used daily by its customers for the next 20 years. That's customer loyalty in a nutshell and loyalty is earned one customer at a time.

The most successful and profitable businesses are those who embrace CRM and have a well-defined loyalty program. If a company can achieve a loyal customer base, it is well on the way to becoming a market leader. A Bain & Company, Inc. (an international strategy consulting firm) study entitled "Making CRM Work" found that companies can boost revenues by as much as 85% if they can increase the retention rate of their best customers by only 5%. A 2001 study published in Harvard Business Review concurs. That study also indicates that a 5% increase in customer loyalty translates into a 25% to 95% percent increase in profitability

The rewards that a company reaps when it woos and wins a loyal customer base are many. The benefits for individual businesses will vary, but here are just a few that pertain to almost any business:

* Reduced sales and marketing costs.
* Lower transaction costs.
* Increase in sales — cross/up-selling (a larger share of each customer and more customer referrals).
* Less price competition.
    A loyal customer base requires less time, effort and money to serve since an

established customer tends to spend more over time. Also, satisfied customers bring in referrals. It's not that difficult to calculate the ROI that results from founding a loyalty program that builds a loyal customer base. Just determine the:

* Cost to acquire a customer versus the cost to retain an existing customer.
* Cost versus profitability ratio for acquiring additional wallet share from an existing customer.
* Influence value of the existing customer base, i.e. the net present value of all future revenue streams from a customer compared to the net present value of the lifetime cost to maintain that customer relationship.

Developing a consistent, seamless and personal approach to the customer creates a long lasting customer relationship, leading to customer loyalty and increasing company profit and success.

An effective CRM system allows a business to speak to its customers with a single and consistent voice — regardless of the point of contact. That's because everyone within the company shares the same customers' transaction history and information. Components within a CRM system also can help to identify the actual cost of acquiring and retaining individual customers. That information lets a business focus its time and resources on its most profitable customers. But still, it's critical that the staff is trained to collect all relevant customer information and insights.

⊙ THE REALITY

No longer can a company assume that because it has strategically located it's retail outlets or that it has an adept field sales staff that its customer base will remain loyal. The average company's customer loyalty quotient is under fire due to a number of reasons.

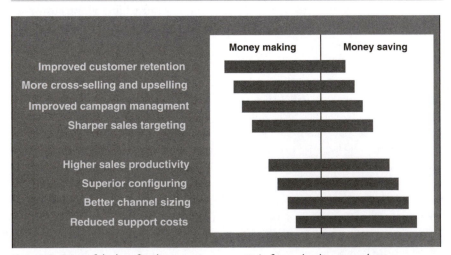

Figure 12. Some of the benefits that a company can attain from a loyal customer base.

The Internet changed everything. Today there is a plethora of information and offers just a mouse click away. This results in a very fragile customer loyalty quotient.

But the Internet also provides ways to increase market share. Many businesses have found that a web strategy can help to increase the customer loyalty quotient, obtain additional sales, and even find new customers; but they've also found that it's not all peaches and cream — just read the daily newspaper.

If a business has a web strategy, it must compliment both the business and the industry as a whole, if it's to be a successful CRM channel. For a perfect example of this type of strategy, visit Office Depot's website (www.officedepot.com).

The customers themselves are different. They are increasingly astute purchasers of goods and services, i.e. they are more demanding about quality, value, convenience and price — simultaneously. This has a huge impact on a business's customer loyalty quotient.

When it comes to customer service, there's no room for error. As the business community tries various differentiation techniques, to not only win the customer, but also to maintain a healthy customer/cost ratio, customers find themselves encountering different levels of service on a daily basis. But the customer is adept at adapting — they are increasingly skilled at assessing customer service and when a company's customer service fails, there is risk of losing the customer. The successful company will cater to this new paradigm.

View the customer as an ally, not a foe. Even when a company does everything right, it still must face razor thin margins and cutthroat competition. Survival in such a climate is a continuous struggle. The business community is no longer in control; it's the customers who have the upper hand. So, work with the customer. Find out their wants and needs. Give accurate information concerning price, delivery date, order and account status.

⊘ THE PROBLEMS

To achieve sustained customer loyalty, look at the company from the outside in, rather than the inside out. The company needs to align its entire operation to deliver service to the customer; not hide behind traditional functional departments and processes.

Differential Difficulties

Businesses seeking to attract and keep customers often find it difficult and expensive to differentiate themselves from their competitors. It is no longer enough to provide different types of service or product, or offer the customer a variety of contact points or channels. Companies must optimize their use of channels, service levels, customer care and customer interface to support their products.

Customer Service

Earning a customer's loyalty has become more difficult. Merely satisfying the customer is not enough to win their loyalty — just because the market research may reveal that the existing customer base is satisfied, doesn't mean the company and its products/services have a competitive edge. And while it's widely touted that poor service is the leading cause of customer attrition, according to a University of Texas study, over 75% of customers who claim to be *satisfied* still show a willingness to wander.

The business community needs to take the extra step; create an environment wherein the customer is enveloped in customer service. Give more value to the customer than the customer has given to the business. This is how real loyalty is won. Only then will customer surveys begin to show such euphoric terms as "very pleased," "extremely satisfied," "excellent," and so forth to describe the customer service experience. The customers who use such emphatic adjectives are found to be six times more likely to repurchase a product.

Inadequate Customer Tracking

Most businesses don't or can't track customer behavior adequately enough to understand their customer loyalty metrics. When analyzing customer loyalty, look at repeat purchase and share of customer statistics, not measurement of customer satisfaction alone — satisfaction is what the customers *say*; loyalty is what they *do*. The business community needs to build a broader view of customer behavior to identify what influences their purchasing decisions. Identifying loyal customers can reduce customer acquisition costs by at least 25% and increase the average order size by more than 50%.

Staples Inc.

Within Staples stellar CRM strategy is a customer loyalty program, which it calls "Customer Dividends." Staples has instituted a world class CRM system, which can parse data about its individual customers — from the large Fortune 500 company to the small SOHO office. It uses this data to create metrics that allows it to determine which current customers will have a propensity toward becoming a lifetime customer given the right incentives. Then it goes after that customer group by inviting those customers to join its Customer Dividend program. The program gives the customer value (a quarterly dividend based on amount purchased during the quarter, free shipping, etc.) and at the same time it helps to give Staples a larger customer loyalty quotient, builds its sales volume, and creates an invaluable lifetime customer relationship.

Wrong Customer Focus

The focus of the business community has been (and still is to some extent) on attracting new customers — offering discounts and other incentives to the first time buyer and compensating employees more for winning a new customer than for keeping an existing customer happy. The focus needs to shift more toward nurturing the existing customer relationship.

A loyal customer base consists of a high percentage of customers with a pattern of repeat purchases and a retention rate that continues to grow. But don't think a high customer retention rate equals loyalty, that requires a high rate of "share of customer" (the ability to increase the proportion of an individual customer's purchases of a company's products and services). This is the same principle upon which Staples Customer Dividend Program was founded.

☼ THE FIX

A good CRM focus turns each interaction with the customer into an opportunity to learn more about that customer. Once the company knows its customer base, it can differentiate itself and its products and services based on that knowledge. This requires capturing every episode of a customer contact by drawing on skills, procedures, and data brought in from a variety of departments, functions and processes. When a company can track all customer interactions it can build a pattern of customer behavior and usage. A company that is able to identify that behavior can then predict and ultimately shape it. But it doesn't end there.

To serve the customer *superbly*, there cannot be any friction in the drawing together of different skills and procedures. The first step in the process is for the company to see itself from the customer's perspective, then make that view as pleasing as possible. This includes every interaction — from the moment the customer's need is first identified to the time when it is finally satisfied. Meaning that all employees who take part in the customer interaction can seamlessly reach across all departments, processes and functions, as needed, to serve that customer.

The next is to provide a unified view of the customer — take all of the information that resides in data silos and make it available to all who need access — customer and employee, alike. Until the business community can adopt such an approach, it will continue to see high failure rates among its members.

A Customer Loyalty Program

To obtain the goal of superb customer service, which leads to a loyal customer base, a company needs to implement what I call "a customer loyalty program." The first phase of that program is to:

* Identify the existing customer base. This must be done before the business can begin to build and retain a relationship with these customers.
* Determine how the company can differentiate itself in the minds of its *existing* customers.
* Clearly define the existing customer service strategy, then identify changes necessary to provide stellar customer service.

* Put into place technology that will give the customer the same access to resources (i.e. there should be complete and consistent information across the channels) and the same service level regardless of the channel used.
* Define the elements of the CRM strategy that embodies the heart of the customer loyalty program.
* Communicate the new customer service strategy to *all* — not only within the company but also to outside service providers. Clearly delineate everyone's responsibilities within this new strategy.
* Develop and utilize training programs to support the new customer service strategy.
* Hire, develop and retain the right people.
* Develop and use quality control strategies to monitor progress and to help "stay the course."
* Institute appropriate reward and recognition programs.

A loyal customer is quick to not only repurchase as the need arises, but to also acquire new products and services as they become available. A higher customer loyalty quotient is the key to strong residual market share and it's effective protection against customer attrition. To achieve the highest quotient possible, a company needs to build service into its overall business strategy. As illustration: A company could institute programs to develop coaching reinforcement skills within the management ranks; and hire or train front line personnel so that they have the technical skills and behavior skills that can deliver service excellence.

Create a Customer-focused Environment

A customer-focused company is characterized by how closely it listens to its customers and then how well it uses that knowledge to provide a standard of service that reflects its customers' wants and needs. The successful business is one that focuses on its customers and listens intently to what they need.

Creating a customer-focused environment begins with the knowledge that loyalty is earned one customer at a time. Why? It's very difficult to achieve loyalty from a group of people. If every customer within a specific group is treated the same, the results will be euphoric praise by a few, satisfaction by some, and disappointment by others.

Establishment of a consistent, positive customer service environment customized for *each* customer is required to establish a loyal customer base. Because individual customers needs and expectations vary widely, it is often necessary to diversify the treatment on a customer-by-customer basis to produce a consistent, positive result. It's not as difficult as it sounds.

Take the example of a computer literate customer, who has access to a computer with Internet access 24 hours a day. That customer is more likely to want to transact most of his or her business via Internet-based channels — email, website, chat. But, if a customer doesn't have access to the Internet or is computer-phobic, then the telephone may be the channel of choice and he or she probably would not appreciate being reminded during every contact that the web is just a click away. Finally, a company may have customers who suffer some type of disability that preclude: use of the web, or in-store visits or use of the telephone, but these customers may find catalogs and regular mail a satisfactory channel to do business with the company. Exploit each individual customer's preference. How? Read the data residing in the customer data storehouses.

Let's say that all of those customers fit the same demographic, geographic and psychographic profile. Most companies would then put them within a specific "customer group" and treat them as one. Unless the company has the means to analyze each customer's channel preferences and purchasing habits, it could place some of the customers at risk when it tries the same customer service techniques for all.

That's not to say there won't be changes that will be advantageous to a specific customer group or even the entire customer base. For instance, a customer becomes irritated when he or she is compelled to explain a problem to various people working in different departments, while being shunted through voice-mail hell. Taking a simple step such as providing an "opt out" to a "live person" can be of benefit. But, taking it one step further and putting the right technology in place where each piece of the puzzle can travel with the customer, providing the next problem solver with the complete picture is true CRM. Or perhaps a customer wants to place an order but at the same time wants to inquire about an outstanding service call, this should be possible without everyone jumping through hoops.

From an internal point of view, employees should have information at their finger tips that let's them know the customer they are dealing with is "special." This alerts them to respond in a unique way.

CAVEAT: Don't become so captivated by technology that the quality of life for the prospective customer or an existing customer is ultimately compromised. As illustration of what NOT to do: Send a new catalog every week. Make too many email offers or telemarketing calls. Pop-up boxes on a website that changes with each page view.

Institute a Cultural Change

Don't discourage customer complaints. Also, don't conceal them. A complaint should be viewed as an opportunity to learn. When a company listens and provides an appropriate response to its customers' complaints, it stops the "bad news travels fast" syndrome in its tracks. In other words, the customer who has been disappointed but made whole through exemplary customer service is less likely to tell friends, neighbors and colleagues of the disappointment. (Alas, the tale of exemplary service is also unlikely to be related.)

Throw out the old internal statistics (how many customer calls were answered, how long they took, how fast a package was delivered, etc.). Institute new customer service measurements that track customers' perceptions of how well calls were handled, complaints responded to, and packages delivered.

Establish a major communication initiative with specific focus on top level management. By doing so a company ensures that its senior managers use the new performance measures, as outlined above, and really do trash the old, statistical ones.

Institute ongoing training and coaching programs for executives, managers and front line employees. This is the best way to increase and enhance customer service company-wide. Now the entire company is in a position to treat its customers in ways that will dramatically exceed their expectations and thus garner customer loyalty.

The People Factor

Every customer transaction requires two components: human and business. In the past (and even now) companies tended to focus on the business aspects of customer service, i.e., providing what the customer orders in the most expedient and efficient manner. The customer suffers disappoint if everything doesn't work smoothly. Even when everything goes like clockwork, it's difficult to build strong customer loyalty by focusing only on the business side of the transaction.

It's the people factor that makes the difference. It's the people who decide to staff call centers, help desks, sales positions. It's people who provide a quick, pleasant (or cranky) response to customer queries and complaints. It's people who provide the training of the personnel so they know how to build a lasting relationship.

Whether the customer's impression is positive or negative, it causes a lasting memory. This directly affects how the customer feels about doing further business with the company. (More than two-thirds of customer defections take place because of poor customer service.)

Managing the people factor in a consistently positive way builds exceptional customer loyalty. For instance, a customer orders 100 widgets in varying sizes and amounts per size but only receives 54 widgets and none in the large size (which the company needed "yesterday"). A call to customer service received the rote answer "they are back-ordered, and will ship as soon as they are received and placed in stock." Not very satisfactory. Another customer places the same order with another company and receives the same deficient shipment. But this time, when the customer calls customer service and explains the problem, the employee can access the pertinent databases and find that another division has the needed widgets on hand. Voila — the customer receives the missing widgets within 48 hours. Exemplary customer service! Now, will the customer remember the deficiency of the shipment or only that the complaint was promptly responded to?

Customers tend to be loyal to companies that win their trust over time via a series of positive events. This means the customer loyalty program must be ongoing. For instance, after the initial push — training and implementation of enabling technology — senior management should have informal meetings, with no departmental managers in attendance, to boost the customer care message. This is also where management can obtain first hand information from the front line — a valuable resource for any decision-maker.

Empowerment

Ensure that the right employee, with the right knowledge and skills, is in the right position, to provide the right service. Then empower them. Give them real decision-making power to provide great customer service — the ability to resolve customer problems during the initial conversation. Also, authorize them to take the time to build a relationship with the customer during the period of contact. But don't limit this empowerment to customer service and call centers, include people across the board.

Accordingly, employees need a wide range of customer contact skills and responses and the authority to use them to handle disparate customer situations effectively. It can be as simple as, for example, having employees that can speak Spanish and is savvy in a particular technology that one group of customers use. But also, the employee should be able to use their own initiative as to when to give discounts, to accept return of defective products or parts, and even to authorize free shipment if the need arises.

Finally, to forestall a backslide toward lackluster customer service, institute a program that provides for sufficient opportunity for individual employee recognition.

Changes in the corporate culture are possible only if the executive suite and the top-level managers are committed to the change process. Half-hearted reform won't get the job done; in reality such half measures may ensure the worst of both a worlds — a business turned 180 degrees internally, but a management team stuck in the old milieu.

❂ BENEFITS

As the level of overall customer satisfaction grows, it can stem the tide of customer attrition. There are also added-value benefits. For example, as a customer-focused environment matures, it can affect sales functions — sales are not only boosted, but have a quality that enriches new and existing customer relationships. All because there is now a well developed strategy to deal with customer contact. The integration of sales and service, and the ensuing corporate synergy, can be the secret weapon of a company with a customer-focused environment.

Armed with such a high level of information, a company can focus on customer retention to reduce churn (for instance through loyalty programs), to increase sales in order to produce a net growth of profitable business, and then to cross-sell further services to existing customers. These are no longer desirable goals. They are imperatives.

Profitability is closely linked with customer loyalty due to three basic impact points:

1. Revenue Growth. A large, loyal customer base means higher profits since customer spending tends to accelerate over time and they also provide a high referral rate.
2. Operating Efficiency. As customers and the company get to know each other, both become more efficient.
3. Price Premium. Price is no longer the competitive differential

✪ IN SUMMARY

CRM should be an outgrowth of a new, customer-focused environment. This means creation of an environment that drives reorganization of the people in the company around its customers, which necessitates re-engineering how people work. While all of this is supported by CRM technology, it's the changes within the human element of the business that ultimately builds a loyal customer base and brings increased profits.

Customer loyalty is the competitive advantage in business today.

Chapter 8
Know Thy Customer

The business community has set its sights on the customer as the key revenue generator for delivering sustained and profitable growth. To hit the bull's eye, businesses must understand what makes their customers tick. Yet, many times the "powers that be," in their rush to implement a CRM initiative, forget that a better understanding of the customer base is needed to build a better relationship with the company's customers.

In simple terms, CRM is serving customers better by knowing them better. Successful businesses always strive for this. Now there are new technological tools that can help to gather and use customer knowledge in smarter, faster and easier ways.

With all the hubbub surrounding CRM, you'd think the customer is a new kid on the block. Instead the customer is more like an old friend who's been taken for granted — at least until now. In years past, businesses used a product-centric approach to control all interactions with their customers. Now, many companies are transitioning from a product-centric to a customer-centric focus — where the customer, not the offering company, defines the rules. The business community is responding by (re)designing its processes, products and services around its knowledge of its customer base.

CRM enables a company to understand what its customers want. The customer makes the rules. For example, customer A orders product X (a design based on the wants, needs and desires of the customer base) "as is." Customer B wants product X with a bit of revision — the addition of components 1, 2 and 3.

Customer C wants product X with components 1 and 3 delivered overnight. Finally, institutional customer D insists on a volume price cut, since it bought 100 units of X.

With rules-based engines running behind the scene, on-the-spot decisions that can affect the outcome of a sale are easy — as in each of the preceding examples, where a customized product is provided based on a real-time individual customer's demand. With Customer A the traditional processes and rules are brought into play — are the products and/or components in stock? When can delivery be made? But for Customer B there is a need for new rules that can provide real-time decisions, such as will the requested components work with product X? What is the price of the new configuration? Is overnight delivery an option and at what cost? Customer D not only needs to know if 100 units are in stock, but at what level of purchase is a price break given (100 units, 125 units, or what) and how much that price break will be.

With a CRM culture and architecture, a company can offer products and services that are dynamic and innovative, constantly validated and measured, and configured to meet the customers' needs. But companies also can listen to their customers. For instance, to create value in the mind of the customer, companies rush to develop *relevant* customer offers — bundling products and services — tailored for an individual customer's needs and wants.

How can a company know what an individual customer wants and needs? By having a central repository for all data and the necessary tools to analyze and mine the data based on defined criteria, such as locale, sex, income, household makeup, annual purchase amount. Then analytical and modeling tools can parse the mined data to identify, segment, and profile the company's customers based on characteristics such as profitability, lifetime revenue potential, and cost to service; as well as interests, affinities, buying habits, and service requirement. This provides real customer knowledge and customer intelligence, which provide the direction for building systems and processes based on the needs of its customer base and for allocating resources toward customer segments that are likely to yield the greatest returns. Then through focused customer surveys, personalization engines, relationship marketing tools, and other means of information gathering and processing, a company can develop a unique, fact-based understanding of the needs and future needs of its individual customers. This is referred to as "customer insight." The company then can draw on this customer insight for direction. Through customer insight a company can focus on crafting compelling value propositions and developing new products and services that are coveted by the individual customer. All of which helps to create strong brands and allows a company to efficiently managing its product/service mix.

To reach the Valhalla of CRM, a company must evaluate its customers' behavior intelligently, on a continuous basis, at all phases of the relationship, i.e. during such phases as brand-awareness, sales-execution, and service and support. This includes providing continuity in the customer experience across all touch points and channels. Most importantly, CRM tools must be used in conjunction with smart management.

⊙ ASSESSING THE CUSTOMER BASE

Prior to implementing a CRM initiative, perform a customer information assessment. Many times, a business is unaware of the sources of customer data available within its own corporate walls. This assessment will help a CRM team to define the information requirements of the CRM initiative, which will, in turn, determine the technology needed to support the CRM initiative. Then, if needed, and in many cases it is, a company can also develop a customer-centric data repository that integrates all customer information from a wide range of sources.

The CRM team should also outline what actions are to be taken in response to the customer information gathered. Customer intelligence is one of the pleasant surprises that awaits at company at the end of an expensive CRM project. Customer intelligence can help to not only ensure the success of a CRM initiative, but also the company's customer relationships, and ultimately, the business, itself.

CRM is all about the customer. Thus, customer analysis is key to any CRM initiative. It's impossible to implement truly effective CRM without intimate knowledge of the customer. Different customers have different priorities — one may want extra service and another may value a faster delivery time.

One of the main functions of CRM is to deliver individualized service in the areas each customer values the most. This allows a company to apply its resources where they are most appreciated and thus, deliver the greatest ROI.

Identity

Before a business can parse the information necessary to treat each customer as an individual, they must categorize their customers by key characteristics, needs and value to the company. Understanding customer segments and key customer characteristics are vital for targeting appropriate products and services and recognizing emerging trends. Some questions to ask when performing data analysis:

* What is the number or percentage of customers by geography?
* What are the characteristics of each customer (company size, revenue, industry or age, income, lifestyle, family size)? Is the customer new or a repeat?
* What number or type of products are purchased by the customers?
* What channels do the customers use?

This information is used to build a customer's demographic profile.

Demographic profiles typically remain consistent, but the customers' expectations change. They change based on recent experiences — online, offline, with competitors, with unrelated companies, etc. Frequent testing is a necessary business process since changes in customer characteristics can signal new risks and/or new opportunities. For example, changes can occasion new or renewed emphasis on different sales channels, selling strategies, marketing campaigns or customer support initiatives.

Loyalty

Although I devoted an entire chapter to customer loyalty, since it's so vital to a company's success, I provide a short review. Understand who's staying and who's leaving and why. Most customers don't decide to stay or leave on a whim. Instead, there are a series of decision points, which, if monitored, can give the company a "heads up." Changes in purchase frequency or amount, usage frequency and duration, payment promptness, and other behaviors are all indications of a customer's loyalty.

Everyone knows customers who use a product on a trial basis must have a positive experience to become a buyer. But, not so well known is that it takes several repeat purchases before a new customer develops the buying "habit."

While canceled orders are a sure sign of a dissatisfied customer; many customers just let their orders gradually diminish until they fade away entirely.

Benefits of Customer Analysis

Analysis tools can identify the company's most valuable customers (i.e. those who are profitable, have a low cost to service metric, and will provide a high lifetime value). These tools can take the analytics further — to determine a customer's profitability metric — separating out the customers who purchase the highest dollar amount, then weighing that number against costs incurred to support them via their chosen channel. For example, such analysis may show that when all data is considered, it costs more to support a customer whose chosen channel is the telephone than a customer who habitually uses email as his or her main source of contact.

Next, through further analysis, the profitable customer segment can be separated into target groups based on buying patterns, product interests, affinities, and myriad other characteristics. Good examples of the result of this kind of analysis are the "gold," "bronze," and "platinum" programs that cater to a loyal cadre of big spenders.

Once customers are segmented into target groups, campaigns can be developed that introduce targeted valued-added information and offers. Many customers appreciate these direct marketing efforts.

There are a variety of analysis, modeling and data mining tools that allow a business to measure loyalty in terms of both the number of customers retained and the amount of revenue generated to see patterns that might otherwise be missed. Many CRM initiatives can benefit from these tools.

Note: It takes a special skillset to use these advanced analytical and modeling tools. Many business executives don't have that skillset. Thus, to get the most out of such tools, the management staff will need to be trained in not only how to use the tools but also development of the tactical, strategical and analytical skills, which must be employed.

Value Quotient

CRM solutions are most effective when focused on the right customers, i.e. the most valuable customers. Surprisingly, the most valuable customer might not be the biggest spender (who may be high maintenance and thus, costly) or the most frequent buyer (frequent orders may mean small orders; i.e. over time very expensive to fulfill).

While a customer's value quotient is difficult to calculate precisely, with the right tools onboard, it can be approximated by tracking discount rates, product profitability, customer acquisition costs, cost of service (the number of support calls, fulfillment costs, etc.) and account management (the number of dedicated or shared resources).

Satisfaction

Customer surveys (to understand key customer expectations) and company service metrics (to get early warnings on potential problem areas) can help determine customer satisfaction levels and problem areas. Pay special attention to company handoff points (such as from the call center to accounts receivable, website to inventory management and so forth). Typical metrics include call-hold time, on-time shipment, order accuracy, product quality, product availability, shopping cart or call queue abandon rates, customer response time, and customer complaints.

The Results

For years the business community has gathered *samplings* of historical customer information and placed it at the disposal of only a select few. Now, with the right technology in place, historical and dynamic transactional data can be made available throughout the company giving access to a very specific view of each customer's needs, behaviors and intentions. This flow of continuous information can drive tactical decisions at the individual employee/customer level, enabling companies to approach a one-to-one customer experience.

⊙ CUSTOMERS' ROLES

Supporting customers in the various "roles" they play is the essence of CRM. Every customer will assume different roles at different times in their relationship with a company. CRM can identify these roles as they relate to a business and support the customers as they move in and out of each role.

The Prospect

One such role is the prospective customer who must be wooed to gain his or her business. (Even existing customers may play this role at various junctures during their relationship with a business.) That's why sales force automation (SFA) software is a major force in most CRM initiatives. SFA applications isolate, manage and channel sales opportunities through the sales pipeline, managing the opportunity from "cradle to close." But for this to occur there must be a fully integrated CRM system, as opposed to a series of information islands.

Probably the greatest sales competition any salesperson faces is not another salesperson, but simply ignorance of the role their customer is currently playing. The sales person is many times the last to know if a customer has donned the prospect role because customer service and support staff are usually the only regular contact a customer has with a business. This employee group is normally the first to hear that a customer is looking for or needs something new or improved. With integration between software supporting customer support centers (call centers, service centers, and help desk) and the sales department's CRM components, customer leads can be quickly channeled to the correct sales person for rapid follow-up. Without integration, opportunities may be lost and customers might go elsewhere.

The Account

For many businesses the bulk of their relationship management is in managing the customer in an account role. When a customer contacts a call center to question a statement, or a help desk for help in getting a product to work correctly, or a service center to get a warranty claim satisfied, specialized software tracks the volume of disparate interactions, creating a very useful database of customer transactions that a company can mine. The data gleaned can be as diverse as what types of customer cause the most hassles, which products and services get the most praise or complaints, which customers need upgrades, training or a visit from a friendly sales person.

Effectively managing the customer in an account role depends on providing a high level of customer service, which is why portal software has become a new addition to the CRM universe. Portals give companies the chance to concentrate much of the customer service experience through a single gateway — a website the customer can access over the Internet. American Express, Con Edison and Verizon are examples of businesses that have built portals for their customers. These portals not only provide self-service functions such as account balances and bill paying, but also access to added-value services (such as purchase of service enhancements and travel bookings). Portals have become one of the key ways a business can personalize and expand the scope of its customer relationships without the need for additional personnel. Portals can even provide one-to-one marketing opportunities.

The Business Partner

When portals link to ERP systems, the customer can take another role, that of a business partner. ERP systems can manage customer order processing workflow, customer billing, customer credit status, collections, customer cash receipts, customer-specific inventory, or service pricing rules.

An ERP system communicates vital bits of information about customers to integrated CRM tools for use in various CRM activities. ERP-CRM integration provides a holistic view of a customer's relationship with the company, i.e., should customers on credit hold still be in the sales pipeline? Will a customer still pay the latest invoice, statement or bill if the service department has stumbled? If an effort is made to promptly notify customers of order delays or back-order status, will it have a positive affect on the customer relationship?

If a company wants to treat its customers as genuine business partners, its ERP system should be accessible to customers via web-based, self-service functions. High-value customers must be able to see anything important about their business relationship as openly as the company's staff. For example, if salespeople know the customer is on credit hold, it should be just as easy for the customer to have the same information via their own secure self-service section within the portal environment.

The E-customer

With a portal or any well-designed website, the customer functions entirely as an e-customer. (E-customers are those with whom a business manages most or all of their relationships electronically over the Internet.) In the past, many businesses acquired customers by actively selling to them via a direct sales or direct marketing process (although occasionally, a business got lucky with a "walk-in" customer).

E-customers are the electronic equivalent of the walk-in customer. What most defines an e-customer is their "attitude," which one writer aptly described as "cruise 'em, bruise 'em, and lose 'em." CRM can help a company manage the customer in the e-customer role by bringing into play a whole new set of technologies that include:

* Multichannel e-customer interaction management (one-click chat, email, FTP, or call-back service).
* E-customer clickstream analysis (capturing and analyzing massive amounts of data about each customer's web session via log files to provide a means to understand an e-customer's behavior and buying patterns).
* E-customer loyalty management (offers and tracks loyalty bonuses gained by frequent shoppers).

The Company Asset

I've saved the best customer role for last — the customer as an asset — a role that all potential customers assume when they become customers. There are many assets in every business — product assets, people assets, equipment assets and cash assets — it's common for a company to use technology to track and manage these assets. But strangely enough, it was only after the arrival of CRM that many businesses realized they had not been actively managing their most important asset — their customers. Increased competition and decreased profitability became the wake-up call.

At its most basic, CRM is a strategic concept that manages this very important asset through management of the customer's interaction with the company. While there are prudent actions that can be taken to protect, grow and assist this asset and thus, improve profitability, there are also foolish actions that can be taken when dealing with this valuable asset, which can harm the business's profitability. For instance, exemplary customer service is a "good thing." But, there should be accountability. How much money should be spent on cultivating a low value customer?

This is where analytical CRM software comes into play. These tools provide templates to allow effective analysis of customer-related data and can even deliver key performance indicators, which are focused on measuring and monitoring the performance of the customer asset. Analytical tools can compile structured knowledge bases from data stored in customer call center databases, email messages, or web log files, as well as through software that continually surveys e-customers with pop-up questionnaires that ask: Why did you come here? Why did you buy this? Why are you leaving the website? How much money have you spent over the last year? What channels did you use?

The business community has come to realize that customers are the only assets that count. A business's worth is defined not only by what the auditors say, but also by what its customers say. In many ways, financial statements reflect the customers' opinion of the company and its products and/or services. Customers are the ultimate arbiters of whether a product or service is worth buying, and they vote with their wallets.

Understand what roles the customers are playing, then aim high. This requires nothing less than a tightly integrated CRM solution. CRM succeeds best when a company presents a single, consistent face to the customer rather than a series of unrelated encounters masking the fact that no central repository exists to manage the customer-company relationship.

⊙ THE CUSTOMER EXPERIENCE

What's involved in providing a complete customer experience, at least from the business's perspective? It's knowledge. Knowledge can provide a company with the customer intelligence and insight necessary to market, sell, deliver, and service its customer base in such a way that everyone is satisfied. The right customer knowledge enables a business to take steps to cultivate a profitable relationship for both the customer and the business.

First, identify the customer and the contact channels used by the customer. Once there is a clear understanding of the contact channels a customer uses, its possible to establish a common view of the customer's activity. This creates a centralized definition of the customer and builds a unified customer experience model, aligning the information the customer might seek to a business's available channels.

Once this foundation is in place, the company is ready to apply analytics to its customer information. A set of metrics are generated that can be used to measure, monitor, and benchmark the customer. Now the company is in the position to apply available resources to the processes supporting that customer.

The first phase is personalization of each customer experience. Many companies have added e-business to its operation model, primarily to provide an additional channel of contact for their customer base. However, while the online community is growing every year, even the most dedicated web surfer will also use other channels to contact a company. For example, a customer may select the self-service convenience of the web in one instance, yet drop a dime (okay, a quarter) to reach a call center for personal assistance in another. Yet, no matter how the contact is made the customer chooses the channel.

The next phase is to optimize all customer interactions. Once the first phase is complete, it's possible to begin the process of influencing customer behavior through marketing, sales, and service and support systems.

The third phase is to leverage the customer base as an asset. Three things are necessary to leverage a company's customer assets. First, the linkage of operational CRM tools with the company's supply chain partners and/or divisions. Second, integration of the customer information gathered in the previous phases with other information such as financial and supply chain data. Third, once parsed, this data will provide a common set of performance and cost metrics.

Finally, develop pricing and revenue analytics that aid in the understanding of variables and actions required to achieve customer profitability. One example is dynamic pricing, such as what the travel industry has been doing for a number of years.

To manage customer-centric processes requires development of processes that can break cultural and business barriers. Analytics drives the operation, optimization, and automation of business operations across all channels of customer interaction.

Using Data Wisely

It's vital that the CRM team understands where customer knowledge, intelligence and insight fit into not only the CRM strategy, but also within the overall business strategy and vision. Once understood, it can be applied to build customer data gathering and analytical processing into the very fabric of the technology infrastructure. This then allows the company to achieve ROI at every stage of the customer relationship building effort.

Let's take a look at a couple of examples:

A business may find through marketing analysis, surveys, buying history, external information, and customer profiles that 20% percent of its customers have multiple potential product interests. It should, therefore, focus on these customers for cross-sell campaigns. On the other hand, if a cross-sell campaign is focused on customers who don't show a propensity toward interest in other products, the results will be limited and may even have a negative result.

Or perhaps, a business that's considering upgrading its web presence may find that only 20% of its customer base uses the web on a regular basis; hence the project is put on hold. However, if it's determined that the current customers are becoming web users in large numbers, making an investment in state-of-the-art web components can provide a high ROI.

Customer knowledge must be kept fresh. It's not a one-time tactic. Establishment of an effective customer relationship is an evolving, learning relationship. The true value will be realized over time as the relationship proceeds to greater degrees of intimacy. This requires continually listening to and learning from the customer.

Don't spoil the relationship by continuously overloading the customer with irrelevant solicitations, offers and information, through mass mailings, extraneous pop-up boxes on a website (my pet peeve), and unfocused sales offers. Coordinate the marketing campaigns. Intelligently segment the customer base into target groups. In other words don't direct a blanket campaign to everyone on the mailing list.

Building and updating a customer knowledge database is the foundation of CRM. A company needs current metrics, at all times, as to what is the average customer revenue, annualized churn rate, current response rate, acquisition cost by channel, and so forth.

To build long-term relationships that are satisfying and profitable for all concerned requires that the business community take the time to regularly determine its customers' needs and review their perception of its performance. Complacency has no place in CRM. If a company thinks its customers are the same today as they were yesterday (and it treats them so), that company might as well order the going out of business signs now.

The total customer experience is what attracts a loyal customer base, locks the customer in, and secures their long-term loyalty.

⊙ THE CUSTOMER'S PERSPECTIVE

Even businesses that rate their CRM initiative a success don't really exploit the full potential of their systems, mostly because they're not approaching CRM with the right mind-set. To realize more business value from CRM, a company must formulate a clear strategy that encompasses three crucial dimensions: business strategy, technology, and the ability for the company to truly understand the customer's perspective. How can a company get a good customer perspective? Have its CRM project team:

* Conduct customer surveys. A company needs to know its customers' wants, needs, desires and disappointments before it can satisfactorily serve its customer base. If a company doesn't take the time to query and listen to its customers, it may find that it has spent money on the wrong functions and features.

* Accompany the sales staff on customer visits. This allows the CRM team to see firsthand the products and level of service the customers are demanding.

* Call customers and interview them on the processes they use to make their final purchase decision. The results can be an eye-opener.

* Take on the role of a customer. Act as undercover agents. Purchase various products and services using different channels. Then have the team place calls requesting support for say, an invalid invoice, a damaged product, a warranty issue, technical support, field service, you get the idea. This is a good way to understand where the current processes are working and where they need improvement.

Some companies implement CRM without fully addressing organizational and behavioral issues. While these companies honestly believe they are customer-centric, from the customer's perspective, they aren't. In reality, all they've done is automate product-centric business practices and paid "lip service" to the issue of establishing a customer-centric environment.

Each independent department or division may believe it is customer-centric because it provides strong customer support. Yet, only when multiple departments share customer databases and/or information and coordinate dealings with customers, is the customer experience seamless. Thus, the company, as a whole, isn't customer-centric from the customer's perspective. Once the CRM team gets a view of the company from a customer's perspective, it will see that as far as the customer is concerned, there are actually a number of different "companies" within the organization. The customer, therefore, evaluates each department or division on its own merits, with one negative sinking the ship.

☼ IN SUMMARY

Ask the right questions, at the right time, using the right methods and media. Audit customer experience through periodic sampling at all customer touch points. This information not only measures CRM effectiveness, but also identifies areas for improvement and helps develop metrics for evaluating and rewarding customer-facing organizations.

Understand the customer base. Find out their needs and priorities. Analyze the company's customer service. How does this service measure up to the customer's needs and desires? Don't stop with internal assessment, look at every-thing from the customer's perspective. Obtain this information through as much personal contact as possible.

Be creative. Use the information gathered to improve the products and/or services offered. Add new offerings, if necessary, and remove those that customers don't want. Think "out of the box." Look at what other industries are doing and adapt them for use within the company's marketplace.

Exceed expectations. Do everything within reason to provide a great customer experience, every time. Make the most of every contact. Each complaint should be viewed as an opportunity. Surprise the customer by taking the unnecessary (but much appreciated) extra step. For example, walk them through a procedure, rather than just referencing them to a self-service website or sending an instruc-tional fax. Give them extra information; ask if there's anything else that can be done to help.

Chapter 9
The Many Flavors of CRM — An Overview

The tsunami of advancements in technology over the last 20 years or so has allowed the business community to take advantage of a growing mass of customer data in increasingly cost-effective and innovative ways. Today's computing infrastructure can handle more volume, enabling businesses to track specific customer information, and to share it throughout the corporate environment. It's now economically possible to store terabytes of data and to analyze that data (enabling a business to offer personalized special services); providing the basis to build and maintain a small town-like rapport with the customer.

Let's not forget the Internet, which has not only provided a completely new communications channel, but has forever changed the characteristics of customer interaction. Just look at the rise in e-CRM applications with web capabilities such as sales configurators, customer self-service functions (FAQs, searchable knowledge bases, ordering and order tracking), and web-enabled call centers (agents communicate via email and chat). Unfortunately, the majority of these e-CRM tools don't provide links to many of the systems that are necessary to provide the customer with a seamless view into the business's customer-facing applications.

The reason for the disconnects is that many within the business community are mired in isolated, outdated, and incompatible technology. Currently, businesses on the whole, use stopgap measures (manual and paper intensive processes) to work around these obstacles. Many have chosen to use point solutions at various junctures in the customer life cycle, to help automate these inefficient processes. While this may offer some advantages for some businesses, especially the small business, these solutions don't have the ability to leverage shared information. Although CRM tools, used independently or jointly, have a beneficial effect on a company's overall efficiency, to achieve the full effect that CRM can bring to a company, there needs to be complete corporate knowledge of all customer interactions at all steps within the CRM process.

❂ THE HISTORY

Before delving further into CRM-related technology, we should first get a historical perspective. In the scheme of things, CRM is still in its infancy — it came riding in on the coattails of the personal computer when personal productivity was the "darling of the moment." What followed was a hodgepodge of products designed to make the individual worker more productive; but these tools had little to do with customer relationship management. These management systems, i.e. call center, marketing, campaign, contact, etc. were each designed to automate only one segment of the customer life cycle. All ignored a fundamental CRM principle — each task making up the customer life cycle is part of an overall process, and the only way to give the customer a seamless view is to link together the tasks to provide a cohesive process.

By the early 1990s this realization had fostered a cornucopia of new technology, all aimed at enabling interdepartmental interaction while increasing personal productivity, i.e. groupware, workflow technology and a crude version of sales force automation (little more than a contact management system).

Together these products represented the first step in breaking down corporate departmental walls. This synergy of integration paid big dividends, and CRM vendors took note. During this period, the main emphasis of CRM was on how to improve internal processes. It was all about automating customer support, sales, and the way in which field sales and service representatives worked with the customer.

Now, while this was going on, engineers were busy building applications to help manage financial assets (SAP and Oracle, for example). Following close behind these applications came technology to help manage other assets, e.g. physical and human assets. Lagging in the rear were the forefathers of CRM — technology for database marketing, call centers and sales force automation. Although I must admit there were a few vendors (e.g. Clarify, Oracle and Vantive) who focused upon automating and standardizing the internal processes of the enterprise such as those in the Fortune 1000. These ranged from capturing sales leads to creating scripts for call center agents. Many enterprises still have such systems in place today.

As recently as 1997, CRM (the term had just recently been coined to describe some rather disparate application areas) was a Cinderella industry. It was overshadowed by ERP (Enterprise Resource Management) and shouldered aside by the media in their rush to cover the nascent Internet phenomena.

Along with the change wrought by the Internet and the web, came an even more jarring transformation within the business and CRM communities. Not only did the ubiquitousness of the Internet mean that customers had another contact channel, it also meant that the client/server architecture supporting existing business and CRM systems might eventually be rendered obsolete.

The Internet wasn't all bad news, it did give the business community a way to track their customers' buying patterns, gave them a new promotional outlet, a new sales channel, and allowed them to get to know their customers better. Sales forces could turn their attention to managing prospects rather than cold calling "suspects" (potential prospects) and call center agents could access a more complete view of a customer's transactions (online and offline).

❂ CURRENT EVENTS

The business community now realizes that its most valuable asset, the customer, has been sorely ignored; ergo the vendor community has taken steps to remedy the situation through technology that can proactively manage this important asset. However, there is some controversy (okay, a lot of controversy) about how this should be done:

* *Data-enabling product-centric processes* — getting the information needed to whomever needs it. But this type of technology presents an efficient versus effective issue, i.e., the focus is still on internal processes and people, and the customer is only secondary. Although this type of technology is often under the CRM umbrella, it isn't CRM; it's more business management. Good examples of such products are Amherst's CustomCommerce, EDS's TeamCenter and Camstar System's InSite.

* *Customer-centric processes* — a more holistic view that focuses on managing processes rather than managing departmental silos. Customer interactions are managed as part of well-defined corporate processes. The emphasis is on effectiveness with efficiency taking the back seat, i.e. the focus is *truly* on the customer. There are many vendors offering such capabilities. A sampling: Applix, Oracle and Remedy.

* The Holy (or perhaps "holey") Grail, adopts a *one-to-one philosophy* as discussed in Chapter 1. One-to-one marketing technologies, such as those provided by e2 Communications are making inroads in areas such as email and data marketing. BroadVision has launched its One-to-One Enterprise product that supposedly employs analytics from another BroadVision product (BroadBase) to support data mart and reporting capabilities, including analysis of content, fulfillment, sales, shopping patterns, visitors, and web activity.

CRM applications that are put into place today (and hereafter) should place the customer at the center of the corporate universe and, if feasible, coordinate multiple business functions so they focus on satisfying the needs of the customer. Many of these products can also coordinate multiple channels of communications — face-to-face, call centers (fax, email, telephone) or the web (chat, self-service, email) — in order to accommodate the customers' preferred channel of interaction.

So how does a CRM team determine what functions a specific CRM tool can handle? How does the team know what part of the operational problem it can handle automatically and how much effort will management need to contribute? How can the company determine what the direct impact of the technology will be on its customer base if the CRM solution doesn't function directly at the touch points? Through education — it's the only way to slug your way through the quagmire of vendors and CRM solutions being offered in today's marketplace.

Some CRM vendors offer truly cutting-edge technology. Others, such as the vendors offering powerful CRM product suites (in the same vein as ERP solutions), promise a bit of everything — integration of CRM with internal, front-office applications, supply chain management systems, ERP and back-office applications. There has also been a recent move to integrate CRM with e-business, since the two are so intertwined.

Whatever CRM initiative being contemplated, to be effective it needs to automate the activities that follow the initial customer contact. This means integrating not only the processes, but also sharing information required downstream; eliminating paperwork, extra handoffs, repeated steps and departmental silos. The results are more satisfied customers, lower operating costs and higher profits.

○ THE CHOICES

CRM technology comes in all shapes, sizes and flavors. It's startling how many vendors have repositioned themselves as CRM vendors — from simple contact management software to enterprise-level e-business suites — everything is fair game. There are hundreds of CRM vendors who claim to contribute vital pieces to the CRM puzzle. If not careful, a CRM team may find that their pieces of the puzzle won't function properly, won't easily blend with other products, or can't provide real-time information. A bad choice can cost precious time and money.

Once the word gets out that a company is looking at CRM technology, it can become a free-for-all. Don't get caught up in the marketing hype ballyhoo. Soon the CRM teams' heads will be reeling with questions such as, "Which CRM?" "What data warehouse engine, data mining tool, campaign manager, ACD, IVR...?" "What function does it provide?" "What will be the impact on the customers if the software doesn't function directly at the customer touch points?" Augh!

The CRM team must know the company's customer base in detail. Then how technology can help the company provide its customers with the services and products they want and need. Different customer relationship approaches, and even different CRM software, may be needed to optimize different types of relationships — one size doesn't fit all.

The number and variety of CRM tools are growing and evolving to meet the needs of the business environments into which they are sold. Still, no matter what the vendor marketing gurus might tell you, there is NO ONE, SINGLE product that can be applied universally. Not Vantive. Not Oracle. Not Peoplesoft. Even CRM-suite vendor E.piphany says that only 70% of its solution is "out-of-the-box." The other 30% requires some level of customization. GartnerGroup states that for the near future less than 10% of CRM vendors will be able to offer a truly integrated suite of products that adequately covers sales, customer service and support as well as database marketing.

Let's bring some substance to the term "CRM." Here are explanations of some of the terms commonly used by the CRM vendor community:

* *Customer interaction management* refers to functions for self-service and assisted service via telephone (landline, VoIP, wireless) and Internet-based chat; customer acquisition sales processes; support of the existing customer base.
* *Customer retention* refers to functions, such as activity at the touch points, proactively offering incentives to revive inactive customer interest.
* Analytic tools can refer to many evaluation functions. But when brought to a mean level, they all refer to parsing the captured data at all customer contact points.
* *Databases* are just what they've always been — software that manages data. But in many instances that data must be consolidated, cleansed and centralized for both efficiency and consistency. This brings to the fore such terms as data warehousing, data marts and data mining tools.

That's just the tip of the iceberg. Since CRM is such a catchall term, there is a need for more specificity in describing exactly what tasks are being optimized and what technology can contribute to the process.

CRM, which historically looked only at the customer interaction area, now includes all transactional data and processing along with the very important role of supply chain and collaborative working. Since today's CRM technology with its analytical, collaboration and multichannel support complements the traditional operational processes involved in supporting all aspects of customer-facing interactions, we now have what are termed as:

* *Operational CRM* — the automation of customer-facing processes. It handles the customer contact and processing. It manages and synchronizes customer interactions in marketing, sales, and service.
* *Analytical CRM* — the use of customer data to create a mutually beneficial relationship between a business and its customers. This analysis, modeling and evaluation helps to optimize information sources for a better understanding of customer behavior, thus enabling the contact to be highly personalized.
* *Collaborative CRM* — makes interaction between a business, its channels and

its customers possible. It provides the means for the customer to contact the company. It enables collaboration between suppliers, partners, and customers, which not only can improve processes, but also can serve to better meet customers' needs.

Each component is dependent on the others. For instance, analytics drives the decision making in operational CRM for the deployment of marketing, sales and customer service processes. But without the data collected via the operational CRM processes, analytical CRM wouldn't have any data to work with. And without collaborative CRM the data processed by the analytical CRM tools couldn't be effectively disbursed and strategic decision making wouldn't occur. Collectively,

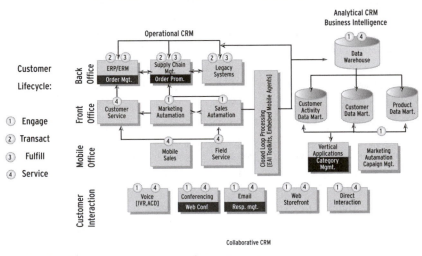

Figure 13. The CRM Ecosystem. Courtesy of Meta Group Inc.

they work in unison to drive the customer life cycle.

In the past the business community concentrated on operational and collaborative tools, but this is rapidly changing. Businesses realize that analytical tools are necessary to drive the strategic and tactical decisions concerning customer acquisition, retention and enhancement. Analytics is what allows a company to *listen* to its customers; without this ability, it's impossible to learn from an existing customer base.

Let's take a closer look at these three categories.

Operational

Operational CRM components automate processes associated with customer interactions, offer multiple contact points for customer communication, and bring efficiencies to customer interactions. Traditional applications include sales force automation; and customer service including email, voice and the web; and marketing tools.

Operational systems:

* Run a business's day-to-day operations and are the primary producers of data. Customer service and call center management systems are prime examples of operational CRM applications.
* Usually require manual interfaces with other operational systems, such as order processing, billing, inventory management, etc.

The output from operational CRM solutions typically are summary level only, showing what activities have occurred, but failing to explain their causes or impact. Unfortunately, this is as far as many CRM initiatives get. While operational CRM does provide business value, by itself it doesn't improve the company's understanding of its customers, nor strengthen its relationship with them.

Analytical

Analytical CRM consists of applications that enable businesses to analyze relevant data in order to achieve a more meaningful and profitable interaction with their customers. It uses customer data for analysis, modeling and evaluation to create a mutually beneficial relationship between a company and its customers, and helps to optimize information sources for a better understanding of customer behavior. Analytical CRM applications include database marketing, sales analysis tools and vertical and application-specific analytic tools.

Analytical systems provide business intelligence. They are the major consumers of the data produced by the operational systems. Analytical systems might consist of:

* Data warehouses, which hold cleansed data that's documented in terms of its source, transformation rules, calculations, etc. The storehouses are the source of integrated, re-engineered, detailed snapshots of customer data.
* Data marts are a subset of the data warehouse. This data has a known set of requirements and reports, and is formatted for a particular function or department. An example might be a marketing department's data mart, where data is aggregated or summarized for use in customer profiling, marketing campaign planning and sales channel analysis.

Analysis is the most critical to CRM success. Analytical CRM solutions enable the effective management of a customer relationship. Only through analysis of customer data can a company begin to understand behaviors, identify buying patterns and trends, and discover causal relationships. Together these help to more accurately model and predict future customer satisfaction and behavior, and lay a quantified foundation for strategic decision making.

The failure to understand the role that analytics plays in the competitive environment can have a growing and significant impact on a business.

Collaborative

Collaborative CRM services and infrastructure make interaction between a company and its channels possible. It enables collaboration between suppliers, partners, and customers to improve processes and meet customers' needs

Collaborative CRM aids a company's internal customer-facing and support staff, mobile sales people, partners, and customers to access, distribute and share customer data/activities.

Collaborative processes are becoming extremely powerful in their ability to tie internal and external faces together to make teamwork easier and more productive. Collaborative CRM solutions provide the means for strategic decisions to be carried out. Business processes and organizational structures are refined based on the improved customer understanding gained through analysis. These solutions allow plans to be revised and integrated across customer-facing activities – including sales, marketing and customer service. They allow businesses to cash in on the valuable insights gained through analysis. Collaborative processes represent the driving force behind today's CRM phenomenon.

Vertical Options

Just as customers assume diverse roles, there are different types of business models with distinct selling, marketing, and customer service and support strategies. Over time different CRM technologies have been developed and customized to support a particular vertical market, i.e.:

* Business models (retail, manufacturing, services, etc.).
* Industries (chemical, banking, clothing etc.).
* Selling and/or marketing strategies (product-centric, database marketing, cold calling, one-to-one, etc.).
* Type of customers (business, consumer, wholesale, retail, etc.).

Just about the only uniform agreement within these diverse markets are that (1) sales, marketing and service are the three areas where the customer makes contact and (2) the customer contact is made in either pre-sales, sales, post sales, or as part of an ongoing relationship. (For more information on industry-specific CRM products visit the "Verticals" section of Chapter 16.)

❂ TECHNOLOGY IN EVOLUTION

The number and varieties of CRM applications are in a never-ending growth cycle with CRM technology continually evolving to meet the needs of the business environment. Even ERP vendors have entered the fray. Their engineers have been working overtime to extend the front-office functionality of their back-office systems so as to provide a full suite of CRM components. They do this by primarily linking their CRM products (e.g. order management and sales force automation) to back-end support. By leveraging their installed base of customers (who also use their database platform), ERP vendors can also tie analytical CRM solutions with their ERP systems.

Today the situation is this: The attitude of most of the CRM vendor community is "this is what we have, work with it." And members of the business community clamor, "we don't care what you have, this is what we want and if you can't supply it, we'll go elsewhere." For a company to deploy the best CRM solution, in a market where evolution is rapid and continuous, it must be able to use the best available technology and to incorporate new or replacement technologies as they arrive on the scene. This means that software that isn't designed to work together must be integrated into a single, seamless service. That requires adding middleware and integrators into the technology mix. (For more on this subject, read Chapter 17.)

Now the decision becomes: Does the CRM team want to go with an established CRM product, then customize it for the company's business model, industry, etc. or go with a probably untried, but specific, vertical tool that is already somewhat customized? Although new to the market, vertical tools *may* offer the best hope since they are more likely to contain the functionality most needed by the company.

Many times, even after the CRM strategy plan is in tip-top shape, the CRM team can still be unclear about which technology will most benefit that strategy. Distinguishing between sales force automation, contact management, lead management, opportunity management can be a chore. Nevertheless, it's essential that everyone on the CRM team learn the basics of these and other CRM enabling technologies.

In the ensuing chapters I discuss in detail some of the diverse technology that falls under the CRM umbrella. All can play a major role in the development, maintenance and enhancement of the customer relationship management paradigm.

✪ SOFTWARE SELECTION

Before beginning the software selection process, the CRM team must have in place an effective CRM vision, a CRM strategy, and a requirements definition for each CRM project. The requirements definition is a document that identifies the business process or rules that make each of the company's CRM projects unique, clearly classifies the business functions which must be automated, and lists the technical features and functionality required. The requirements gathering stage is critical. CRM teams that take the time to gather requirements properly can more easily and quickly select the necessary CRM technology. Neglect the requirements gathering stage and the company will pay in terms of wasted time, effort and money. (There's more information on "requirements definition" in Chapter 16.)

Note: Most CRM software solutions target either the enterprise (500+ employees), the midsize company (100-500 employees) or the small business (less than 100 employees). Some solutions that were initially designed for the enterprise have been scaled down to fit the budget and simpler needs of the small to mid-sized business, i.e. rapid implementation, ease of use and low cost of ownership. However, a system designed from the ground up for the small to mid-sized market is more likely to best suit the requirements of the less than enterprise-sized company.

Focus on the requirements identified by the CRM team as a "must" and then add-on optional features that may benefit the company. Further, whatever technology is chosen, it must be customizable, open, and capable of integrating with existing infrastructure. Give additional consideration to:

* Multi-platform database support.
* Simultaneous data synchronization across multiple database platforms.
* Multi-level security.
* Multi-modal access via multichannel communications.
* Scalability.
* Real-time integration with other applications.
* ERP integration.
* Workflow management tools.
* Analytical engines.
* Web capabilities.
* Easy user-based customization.

With the bewildering array of CRM technology, the CRM team is in for an extensive selection process and that process is anything but straightforward. For instance, many of the best CRM tools are point solutions with their own proprietary interface standards and systems integration issues. At the center of most is the customer data storehouse, but there may also be other core processes, such as data mining, a rules engine and a workflow manager.

Note: *If the company already has an ERP system or a supply chain management system, talk to the vendor. The company may be able to leverage the investment in those systems with a CRM package that is compatible.*

It's possible to reduce some problems if the budget can accommodate a CRM suite package. Yet, often a CRM suite (and many other CRM software packages) will offer modules that aren't needed and lack modules that are necessary. The consistency of quality and usability will also vary between components offered through a suite package. (The following chapters provide more detail on selection criteria.)

⚙ HOSTED SOLUTIONS

The increasing availability of new web-based tools has helped to accelerate the growth of web-based CRM systems, which are commonly referred to as hosted solutions. Hosted solution providers (HSPs) provide a variety of services and as CRM software becomes increasingly complex, and in-house IT staff more in demand, the HSP model may be an attractive alternative. Many are now offering best-of-breed CRM solutions. Also, since many HSPs offer CRM tools at a reasonable cost, smaller businesses may find that a hosted solution offers "more bang for the buck." (I've devoted a complete chapter (14) to the hosted solution model.)

○ IN SUMMARY

The primary catalysts that moved the business community from a product-centric view of the world to a customer-centric one are the increase in global competition and commoditization of products, which make it harder for a business to differentiate itself and its products from the competition. At the same time technology ripened to the point where it was possible to gather enterprise-wide customer information and put it into a single system. It was in this climate that CRM found its place.

A well-constructed and supported CRM vision and strategy is needed. The technology that enables CRM is often *wrongly* envisioned as being able to automate processes with predetermined outcome. *The customer is a wild card.* CRM processes aren't akin to accounting processes, instead they enable a strategy through technical processes and as such, they provide fluid, ever changing metrics that allow a business to *proactively* give the customers what they want and to sell products and services the way the customer wants to buy it.

The CRM field is awash with vendors in all categories. Traditional vendor categories no longer fit and the accepted boundaries between hardware, software and service vendors have blurred. The ensuing chapters represent the author's attempt to inject order into this chaotic picture by concentrating on the technology components that work within a CRM strategy.

Just keep in mind that any viable CRM initiative is dynamic — as one segment is completed, the next starts (hopefully enhancing completed segments). At what point the CRM roll-out begins is often dependent upon the department with the most need. For the purposes of this book, the typical CRM technology-base assistance most companies first adopt usually consists of marketing automation, technology-assisted selling and technology-driven support. Subsets within all of this technology-based assistance are tools specific to data mining, the web; mobile and/or wireless; help desks and field support services.

Chapter 10
Marketing

The marketing department's role is that of a customer advocate. It's the department that spearheads the creation of a mutually beneficial buyer-seller relationship — the tenet of CRM.

A marketing department's mandate is to gather data (through various methods) to help find answers to the age-old questions:

* Who are the customers?
* What is the target market?
* How to nurture a customer relationship?
* How to develop trust?
* How to develop an interactive dialog with the customer?

Then the marketing department is charged with researching and analyzing the amassed data to determine what the targeted customer wants and needs. Based on the results of these efforts, the marketing department can develop campaigns which can direct channel selection, advertising, and promotions; provide leads and other data to sales; and even have a hand in product development and pricing.

✪ CRM TO THE RESCUE

The demands placed on the marketing department are growing, especially with direct interactive marketing; the flexibility of e-marketing; multichannel integration; and real-time dialogue being in almost universal demand. Moreover, new channels, such as wireless (and who knows what's coming down the pike) must be addressed sooner, rather than later.

CRM tools make the marketing department's tasks easier. The CRM vendor community has developed more intelligent, efficient and effective solutions than were available in the past to help resolve many of the issues the average marketing professional grapples with daily.

In recent years a number of techniques have emerged that are designed to improve the relationship between the business community and current and potential customers. The use of such techniques and their enabling technology fall under the general category of CRM and provide features, such as:

* Campaign management.
* Complete contact-history tracking.
* Follow-up management.
* Integrated scheduling.
* Opportunity management.
* Document management.
* Sales reporting and pipeline analysis.
* Budget and forecasting.
* Database marketing.
* Easy data entry.
* Synchronization.

When properly executed, CRM has the potential to create a dynamic marketing system. Still, great marketing decisions aren't made on quantitative data alone, it's how the department parses a divers mass of data derived from both traditional sources and new sources and then uses the resulting statistics to develop a marketing campaign that separates the wheat from the chaff.

How It Works

To achieve the level of analysis necessary for a top notch targeted campaign requires a combination of technologies, including data warehousing to store the

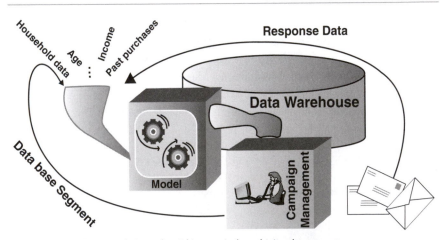

Figure 14. How CRM analysis works within a typical marketing department.

information and OLAP (on-line analytical processing) and data mining for segmentation and understanding of the information. With a comprehensive set of analytical and campaign management tools that support the entire life cycle of a campaign it's possible to:

* Generate leads and pass the information on to sales.
* Analyze and finalize campaigns that will result in the best ROI.
* Coordinate and execute campaigns across all media to maximize penetration.
* Parse customer responses to offer timely, targeted messages about the right products, to the best prospects, using the most effective medium.
* Provide improved ROI, more responses and increased sales.

Campaigns can be targeted to the general population, current customer base or specific customers. There are specialized applications to help manage these campaigns. Many utilize sophisticated analytical tools that can mine the gold stored in customer data that websites, direct interactive marketing, outside sources and "the usual suspects" provide.

Campaign Management

The segmentation of customers and the execution of marketing campaigns define campaign management. As such, it revolves around such items as targeting campaigns; marketing budget management; ad management and placement; and response management. The CRM software components known as "campaign management tools" collects data and automates many of these management functions.

When data is available in real-time, it can instantly reflect changing customer behaviors and demographics, making it possible to obtain richer analysis capabilities, resulting in better segmentation and greater targeting capabilities. This results in prices, offerings and campaigns that are fluid and dynamic, but at the same time easily modified. With sophisticated CRM tools in charge, back-office systems are based on more stable rules and front-office systems are based on ever-changing customer data.

When considering a marketing automation solution, insist that the vendor's sales team demonstrate the product under real world conditions. Then, and only then, can it be determined if the product can:

* Enable meaningful segmentation.
* Provide a productive user interface and environment.
* Manage the complete cycle of a campaign.

The CRM team then needs to interview members of the marketing department to determine if the product meets the necessary criteria. Next, the team should grill the vendor and call its references. (See Chapter 16 for a complete discussion of the vendor selection process.) In other words, determine if the product delivers the requisite "one-two punch" — a fit with the company and its marketing strategy *and* exceptional campaign design and management tools while still providing the features necessary for an end user to run campaigns easily, quickly and with a minimum of training.

Note: CRM can empower marketing departments only after their staff are trained to be strategic thinkers. They are the ones who make decisions based on a complex number of changing variables — product, pricing, competition, customer data and channels — not the software. Looking beyond the rules and applying creativity to the CRM process is the only way marketing professionals can REALLY understand the customer's needs and buying behaviors.

Coordination is Key

Coordinate a marketing-based CRM initiative with other departments. Much of the data gathered is time sensitive, causing the rate of accuracy to diminish if cumbersome transfers, re-keying and reformating of data during hand-offs, and re-processing are necessary before the information can be used. Departmental silos must be breached. As Illustration:

The marketing department must give logistics and fulfillment a heads up so there is enough inventory on-hand and personnel available to handle any spurt in demand that a marketing initiative may create. Consider this situation: In late February, in addition to traditional sales channels, a sales item (say, winter jackets) is posted on a website. Now what was perceived to be a sales device to move leftover winter jackets, causes a landslide of orders from a global base of customers, leaving many potential customers unhappy because there wasn't enough inventory to support such a wide-spread sale. Avoid this situation by either:

* Giving the warehouse advance notice so arrangements can be made to have additional merchandise waiting in the wings.
* Not posting the sale on the website (to prevent the global reach where many potential customers are still in the midst of winter) and removing any mention of the jacket at the pre-sale price from the website (to avoid customer confusion).
* Posting the sale item on the website *and* arranging for a customer pleasing cross-sell item when stock runs out and posting an out-of-stock notice the moment a certain inventory threshold is reached.

In a similar vein, if a marketing campaign has offered a free PDA with every computer order, the warehouse needs to be aware of the campaign so it can have the PDAs on hand.

The marketing department must also keep call centers abreast of all campaigns. If left out of the loop or if the information is inaccurate or untimely, customer service will be inconsistent. The following is a true story. It demonstrates what can happen when there is a lack of interaction and communication between a marketing department and a customer call center.

A call center, handling Los Angeles' metropolitan transportation customer service, opened one morning to an unprecedented volume of callers requesting information about the special fares being offered by the metropolitan bus line. Not one person in the call center had received notice or information about a promotion — at least not until agents reading the paper on their morning break found a full-page ad in the Los Angeles Times! The center, therefore, was understaffed to meet the demand. Thus, the potential of a good ad promotion was lost amid confusion.

Another reason for keeping the web team in the loop is to provide consistent messaging to the customer. For instance, if a campaign offers a two-for-one sale and the website is ignorant of the campaign, the customer is confused. This ups call volume to the call center increasing the cost of the overall campaign to unacceptable levels.

Share marketing's research, written material, etc. with sales. Although the marketing department initiates the sales cycle by planning and running the campaigns, the end results — profiles of the target customers and a list of potential customers (leads) is "thrown over the wall" to the sales department. The sales department then uses various CRM-related management techniques and automation technology to coordinate these leads and sales calls.

✪ MARKETING AUTOMATION

The CRM tools that provide marketing automation vary greatly, and are usually loaded with a wide variety of options. For instance, some can model results, others interface with statistical packages, still others create files for analysis elsewhere. There are so many vendors that claim so much that the end result can be confusion.

Marketing automation refers to automation of the processes necessary to design, execute and measure marketing campaigns:

* Selecting and segmenting customers in a meaningful way.
* Tracking contacts made with customers.
* Measuring the results of customer contacts.
* Modeling campaign results.
* Supporting all campaign media including mail, email, telephone and fax.
* Compatibility, i.e. data can be output and emailed directly to mailing houses for use without rekeying, reprocessing, reformating, etc.
* List management including sophisticated features such as import, merge-purge and deduplication routines.

When considering marketing tools offered by various vendors, only the products that meet the following minimum criteria should make the short list.

Does it:

* Provide a product and price configurator or has it the ability to access one?

✳ Include access to a marketing encyclopedia?

✳ Provide literature fulfillment functionality or have the ability to access same from a third party source?

✳ Offer campaign or project management for both the individual marketing professional and a team effort?

✳ Does it include the ability to create a multi-step campaign?

✳ Measure the effectiveness of a campaign?

✳ Provide overall management of customer loyalty programs?

✳ Offer planning and management tools for special events?

✳ Include linkage to third party applications such as word processors, email applications, spreadsheets, and so forth?

✳ Provide predictive modeling tools or access to same?

✳ Provide mass mailing functions, such as mail merge (can be via access to third party programs) as well as tracking capabilities?

The general marketing tactics currently used require an initial set of mass marketing activities as a first contact ploy, which are followed by more specific campaigns focused on target audiences.

Marketing activities are shifting from traditional telemarketing to web and email campaigns. These marketing tactics are less costly and more appreciated by suspects, prospects and customers, alike. Internet-based campaigns provide a better customer experience since the recipients can retrieve the information on their own terms in their own time. Although to obtain maximum value, follow-up may be needed to separate out qualified leads and provide success/failure analysis.

More advanced software provides the ability to establish, monitor and modify marketing campaigns across multiple channels. These applications can even distinguish between the different marketing channels, allowing marketing professionals to determine the most effective way to approach customers with new products and services.

The most sophisticated software can support different types of campaigns, including single, multi-step and event-triggered, allowing movement from one to another, as circumstances dictate. These marketing tools can also create ROI models, so that the potential outcome of a specific campaign can be calculated and acted upon. Once a campaign has been established, the best match between communications methods and a customer's requirements are possible.

Note: *Shrewd customers expect some level of personalization; therefore, various one-to-one marketing techniques are emerging to better address customers' specific needs.*

The Sales Connection

CRM-related marketing components provide a comprehensive framework for the design, execution and evaluation of marketing campaigns and other related activities, such as interaction with the sales department. Qualified sales leads are one of the typical results of a successful marketing campaign. Thus, marketing and sales automation tools should be complementary and integrated to communicate and coordinate across all channels, departmental walls and systems. See the next chapter for a detailed discussion of the CRM tools available to aid and enhance the sales process.

Database Marketing

Most, if not all, CRM-related marketing tools require the collection and dissemination of considerable amounts of data, which, in turn, requires dedicated marketing databases that can be queried and mined. The resulting nuggets of information can provide the marketing department with detailed analysis, allowing marketing offers to be tailored to match a specific target customer.

Marketing-related CRM technology, for the most part, has its roots in database marketing. Even the most basic marketing automation tool provides a means to segment the customer base and then set that information against sales data. This allows the profiling of customers according to specific criteria. For a database to support a claim to CRM kinship, it must perform the following functions.

* Gather information on individuals.
* Gather inside corporate information.
* Access both sets of information for the development of a marketing strategy.
* Segment and organize the gathered information.
* Generate reports and provide analysis while interfacing with existing applications such as ODBC.

Database marketing, building on the functionality of core CRM tools, allows the use of highly sophisticated statistical techniques, such as predictive modeling, visualization and regression analysis. But first, all marketing-related data must be consolidated into a single marketing database. This requires access to all data points within the corporate environment, then rationalizing them for marketing's use. It also requires data extraction tools, which create and load a multidimensional database.

Look for real-time *synchronization* capabilities in database marketing products. Synchronization ensures that up-to-date customer information is available wherever needed, regardless of geographical or corporate location. It allows database additions and updates to be easily transferred between a central database and a remote database. This enables remote users to exchange up-to-date customer information while maintaining the currency of the information for other users. Shared information can be wide-ranging and can include documents, plans, meeting notes, schedules, forecasts and any other customer-based information stored in integrated systems. The main item to look for in tools with synchronization capabilities is field-level rather than record-level synchronization. Field-level synchronization minimizes data transfer volume, and therefore, reduces transfer time.

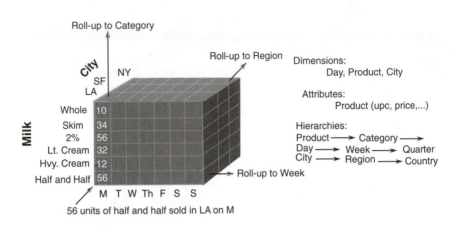

Figure 15. A multidimensional database allows businesses to analyze and match data from numerous different perspectives.

CAVEAT: *Funding a CRM-related marketing program with sophisticated marketing tools with capabilities, such as relationship management, requires sometimes painful and somewhat risky budget reallocations. It's necessary to measure the ROI of these solutions relative to all other marketing assets. Then demonstrate the migration in customer preferences through analysis of their buying behavior. At the same time it must be stressed that if resources necessary to understand, test, and deploy these new approaches aren't made available, the company may miss major shifts in customer behavior and suffer higher selling costs.*

Direct Marketing

For decades, direct marketing programs have offered better targeting, greater personalization, and more efficient use of marketing dollars. Direct marketing delivers the number-one marketing imperative — improved customer relationships. Today's technological tools and sophisticated marketing tactics provide the ability to deliver intelligent, targeted marketing campaigns that can render quality customer acquisition, retention, and penetration. A direct marketing automation tool can provide examples of suspects, prospects and/or customers who meet specific criteria. Such data is very useful when developing a new campaign.

Direct marketing automation products should be scalable and flexible, with the ability to:

* Capture and store detailed information regarding suspects, prospects and customers (the contact).
* Provide the means to develop a precise picture of each contact's value and needs.
* Provide every staff member, who has a relationship with the contact, with an integrated view of that contact.

✳ Identify and differentiate contacts and then modify interaction and services accordingly.

✳ Provide ease of customization and use.

Other features that are desirable include a distributed database approach, and the ability to interface with any ODBC compliant application.

Interactive Direct Marketing

Businesses should pay close attention to interactive direct marketing tools because they have the potential to significantly improve sales and marketing performance and to provide the ability to anticipate and influence the way customers will buy in the future.

Interactive direct marketing tools and strategies, such as targeted web marketing, email marketing, and online customer behavior analysis, provide improved marketing economy and a better understanding of customer buying behavior. These marketing tools focus on the most profitable customers, prospects and suspects. Therefore, they can deliver better performance with more measurable business impact, because they can return more value for the marketing dollar through dramatically improved campaign response rates, and reduction in promotion production and delivery costs. Businesses that aggressively deploy interactive direct marketing tools and strategies collect better information about how their customers buy and behave, which can provide a competitive edge.

Analytical tools can generate behavioral profiles, purchase history, and demographic data, which can help to uncover and anticipate critical customer buying behavior trends that can be exploited to gain competitive advantages such as:

✳ A preference for a specific touch point, i.e. website, kiosk/ATM, call center, or retail outlet.

✳ An increased use of automated processes to help speed and simplify the buying process, such as IVRs in call centers, automatic buying agents for websites, and kiosks/ATMs at brick-and-mortar locations.

✳ An increased reliance on, or willingness to use, trading communities.

✳ A willingness to rely on an automated replenishment system.

Currently the benefits of email marketing and online marketing are among the leading forces behind the move to marketing automation.

Email Marketing

When used with permission-based rules, email marketing campaigns are great. They're the best method for tracking customer behavior. Email marketing campaigns can be more efficient and effective than traditional direct marketing programs, which use regular mail and telemarketing. Direct marketing industry statistics indicate that email can be ten times more efficient than the traditional direct marketing program and twenty times more effective than banner advertising. More impressive — email marketing only costs pennies to deliver. When properly executed, using permission marketing methods, email marketing can cost-effectively build not only customer relationships, but also the overall customer base. Most current permission-based email marketing models maximize performance and personalization and are sensitive to privacy issues.

Technical Considerations

Interactive direct marketing campaigns have specific infrastructure, process, and capacity needs, requiring a sizable investment in CRM-related technology.

For instance, a data warehouse is needed to organize and house a large volume of customer data. Enterprise marketing automation tools turn the data into valuable relationship marketing programs. And finally, integration of call centers (with their own automated email response management and web-based e-care tools) that will manage and respond to millions of inbound email and web inquiries.

E-Marketing

The Internet and e-commerce have sparked a new marketing management category — e-marketing. Operating within this category are targeting tools and ad management services that improve the level of targeting, effectiveness, and customer utility of web marketing programs. These include:

* Online incentive programs, which dramatically improve response rates.
* Targeted web advertising, which can pinpoint ads based on interests and geography.
* Affiliated marketing, which can help businesses expand quickly into new markets by building large reseller networks of virtual business partners (aka affiliates).

Many e-marketing products are increasingly providing components that can create and manage web-centric encyclopedias and knowledge bases. Of course, such components must work in tandem with other components, such as data warehouses and data mining tools.

E-marketing provides a far deeper level of customer understanding than is possible in traditional marketing activities. For instance, monitoring the way in which customers behave when interacting with commercial websites via "clickstream" analysis tools permits a business to track its web visitors' travel patterns as they traverse the site. Once the resultant data is mined (based on specific rules), it's possible to deduce aggregate information about the most popular products and services.

Online customer behavior analysis is the "how, who, what, when, where, and why" of e-business. Today, marketing departments are in a better position than ever to access information about how customers buy on the web. Powerful tools that can turn information into personalized cross-sell and up-sell marketing programs are the key. These tools use profiling services that provide detailed information about customer purchasing behavior as well as personalization and segmentation tools to analyze this behavior. Consider this author-related scenario:

Amazon uses the customer buying behavior data it has collected to offer the author specific books on telecommunications. This offer is based on both my purchase record (I bought a couple of books on broadband technology) and statistical correlation from other shoppers who had also bought broadband technology books or other telecom related books). Although the information presented was of interest, it wasn't topical. Yet, I do believe that over time as the system "listens and learns," Amazon will be able to present me with books that whet my interest.

It's also possible to build a better understanding of the effectiveness of a website, by tracing the paths which users take and where they exit. For example, if an unexpected number of visitors exit from a catalog page with hunting vests displayed, it's usually a sign of problems within the page; or perhaps, it indicates that the vests are being sold on a site that attracts an anti-hunting contingent (presenting a different set of problems).

E-commerce is also plagued with a high level of shopping cart abandonment. It's crucial to understand why a high proportion of visitors begin a purchasing process, yet never complete it. E-marketing tools can provide insight into this conundrum.

Personalization

If the company has a web presence it may consider installing a personalization tool. Personalization can greatly improve productivity and usability while providing key marketing advantages. For example, a personalization tool can select the best offer for each point of contact (whether web-based or via a traditional call center) based on what it knows about a particular customer. These tools are a marvel. When a personalization engine is tied to relationship marketing tools, it's possible to notch personalized marketing up a bit. Personalization tools gather and parse information about a website visitor's individual preferences, so that when a visitor returns to the site their preferences can be automatically loaded. However, for these tools to do their work, customers must cooperate. This requires trust that the company will not share or use their personal information for nefarious reasons.

If the company chooses to implement a personalization tool, don't overlook the privacy issues. Personalization has created a lot of customer privacy and rights discussions and there is mounting opposition to collecting detailed customer information. Reassure the customer that the information that the company gathers will not be misused or sold to third parties. A *well-crafted privacy statement* can address this issue.

A personalization engine is also useful if a business is contemplating a mobile online interface. With a personalization engine running in the background only optimal content will be fed to the mobile customer. This serves to enhance the viewing of content on the small screen of the typical mobile device.

When implementing a personalization tool, always offer an escape (an opt out) from the personalization process. If possible, allow the customer to view and change their personal profile after it has been entered. (Note that existing software can't fully handle this feature — yet.)

Although in general, if a company is selling its products and/or services over the Internet, it will use other types of applications, such as those found in many marketing automation suite products, rather than a personalization tool. These products, which are usually referred to as *relationship marketing tools*, are found in many marketing packages. They can either learn about a customer's preferences over a period of time, or place each customer in a particular market segment. Then the marketing department can use that data to personalize the web experience.

Relationship marketing tools incrementally gather intelligence about the customer by asking the customers to give, for example, their sex, age, income range, and general geographic location. Some, with the right incentives can go further, such as asking questions about financial transactions, health, household makeup, etc. Then the information is used to create personalized automated marketing campaigns that can be subtly introduced to the customers as they surf a website. It's even possible to incorporate other touch points as part of an overall effort to ensure that no sales opportunity is lost. For instance, if a questionnaire indicates that a specific cus-

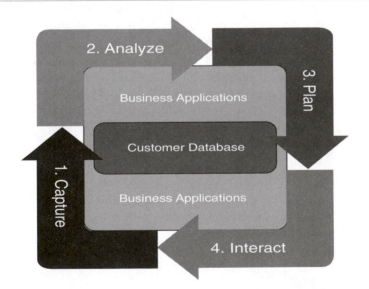

Figure 16. A Closed Loop Process. Customer data winds its way through different systems, processes and databases, generating new actions that are based on dynamic customer feedback, and thus being continually refined to provide improved customer relationships. Through automation these processes and systems can be aligned more closely with front- and back-office processes and systems to create a true, 360 degree view of the customer.

tomer regularly orders from the company's mail order division, information about any outstanding mail orders could be included in that customer's surfing experience. J.C. Penney's website is a good example of this approach.

Another growing trend is *permission-based marketing*, which provides new offers on a permission basis, i.e. the customer is offered an idea, but receives the appropriate detailed information only if the invitation is accepted. No matter the outcome ("Yes, send me information," "No, go away," or whatever) the e-marketing engine is still chugging along, providing feedback to the marketing system about marketing effectiveness. However, be careful that the customer is not bombarded with junk, whether via email, regular mail or pop-up boxes on the website, itself.

Closed-loop Systems

Marketing automation allows a business to present to its audience more timely and more relevant marketing messages that are based on information from multiple systems and delivered across multiple channels, as part of a closed-loop system.

Some marketing automation products focus on the Internet. These tools leverage the unique characteristics of the Internet in ways that can enhance rather than impair marketing efforts in other media. However, the e-customer must be targeted carefully since the Internet attracts customers with different characteristics and preferences than those that populate more traditional channels.

○ FINDING THE RIGHT FIT

How does the CRM team choose the right software and then integrate it into the company's infrastructure? The first step in finding the best technical tools to fit the company's specific business requirements is to meet with the company's IT staff, the marketing personnel, and any other end users (sales and call center personnel) to hammer out details such as:

* How platform, system and database requirements are to be balanced against the number of suspects, prospects and customers from whom information will be gathered.
* Whether there are existing data storehouses, systems or software, and if so, how best to use them.
* What are the goals and objectives of the marketing department's CRM project?
* What are the desired features and functionalities? How will these further the marketing department's goals and objectives?
* Even mundane issues like browser preferences should be discussed.

The second step is to focus on understanding the corporate culture. Where is it now and where will it be in six months, one year, five years, etc.?

The third step is to look at the marketing strategy currently in place and the marketing strategy that will be in place after marketing is automated.

Then, and only then, is the CRM team in a position to determine which competencies, skills and technology will be needed to deliver the marketing messages effectively and profitably. Specifically, the right marketing automation tools to best help the company achieve a competitive advantage.

Once these issues have been resolved, the selection of marketing automation tool(s) that can address all issues can go forward.

A best-of-breed marketing automation system will deliver all the customer information needed to achieve greater ROI from CRM technology by enriching customer data with attitudes, awareness and preferences and then seamlessly integrating it with the overall CRM solution to:

* Build complete customer profiles, bringing together information about behaviors and attitudes from all touch points.
* Collect, consolidate and analyze all the data to make informed predictions about what customers will want next.
* Segment and target customers to offer them the products and/or services they're most likely to purchase.
* Gather the right information and deliver it in time to make effective, strategic marketing decisions.
* Better predict future buying behavior with a comprehensive, integrated customer view that provides information from all touch points, including data on customer buying preferences.
* Reduce costs for online market research by automating processes for data gathering and reporting.
* Reduce costs for ad hoc market research with continuous "event-driven" data collection at digital touch points.
* Improve timeliness, accuracy and completeness of customer feedback with customized online survey data that is bridged to internal data storehouses.
* Respond to customer preferences faster.

No two companies have exactly the same marketing issues (even Pepsi and Coca-cola); functionality needs differ when it comes to automating marketing procedures. But at the same time, commonality, such as workflow, execution (delivery of an outbound message to a segment or target over a specific channel), personalization, response measurement and response modeling, require that priorities be set. Only after there is a clear sense of priorities, can sensible tradeoffs be made — all CRM tools have business tradeoffs including marketing automation tools.

While the vast majority of marketing departments would like the ability to automate all of their processes, most will defer or forego at least one or two due to cost issues. For the small to mid-sized companies the decisions are usually based more on costs than on needs. For example, the tools to enable personalization, response measurement and response modeling are usually beyond this group's budget.

☼ SOMETHING OLD, SOMETHING NEW

Marketing executives should have a good understanding of the CRM-related tools and tactics available to them. Otherwise they won't be able to capitalize on the potential of these new, wondrous tools. Or they will ineffectively manage the many risks posed by these tools, such as rapid technical evolution, industry consolidation, privacy, and marketing infrastructure issues.

For many marketing executives, the challenge is deciding how fast to make the change. On the upside, action taken immediately can capitalize on the potential of a new generation of tools and strategies. These tools have the potential to significantly improve sales and marketing performance and better forecast future market behavior.

On the downside, unnecessary risks may be involved in moving too quickly. Care must be taken when a move is made to abandon proven marketing formulas that generate predictable sales and satisfied mainstream customers.

So, determine the best combination of "new and old" to meet the company's marketing objectives. This may require a move out of the "comfort zone" and some experimenting to assess how fast customer behavior changes and how aggressively to adopt new marketing tools and techniques.

Also, bear in mind that to successfully manage, for example, an interactive direct marketing campaign requires a different mindset, management skillset, and discipline than managing a traditional campaign. This may require that the marketing department's staff be re-educated to get the most from these marvelous products.

Executives and department heads need to develop an understanding of which automation tools and marketing tactics have demonstrated the greatest business impact in their industry. Before hitching the company's wagon to any set of CRM-related marketing tools or tactics, test their performance and ability to segment customers based on current and future buying behavior.

There are many products available — most are feature packed and highly configurable. The Catch 22 is that the more feature packed and configurable the product is, the longer it will take before the system is usable, and an even longer period is required before there is efficiency in its use.

☼ BENEFITS

It's difficult to measure the ROI for any marketing automation project since most benefits are intangible rather than tangible. However, here are a few facts and figures that the reader can use when trying to drum up support for a marketing automation project.

Aberdeen Research (a Boston-based computer industry market research, analysis and consulting organization) found that from 1999 to 2000, email marketing grew by more than 270% and has emerged as the killer application for marketing. Aberdeen"s research also indicates that email marketing will continue to grow through 2003, based on its simplicity, cost-effectiveness, and ability to retain and cultivate long-term customer relationships.

Industry publications indicate that many companies have found that a marketing automation product with features that segment customers in a more sophisticated way can contribute to greater retention and lifetime value benefits. As illustration:

A Canadian hotel chain, Delta Hotels, increased repeat business by leveraging marketing automation tools that segment and target certain customers with an above-average interest in a specific marketing offer. Delta saw a 10% increase in room nights from privilege members in 2000 compared with 1999. A Delta spokesperson told *InternetWeek*, "I'm still evaluating all the data over the past year to see how much Delta can attribute to various initiatives over the past year using the application. Is the increase because we enticed customers to stay at a Delta hotel? I'd like to think so. I'm not ready to say that for sure."

Putnam Investments, a Boston mutual fund company, found that its E.piphany marketing automation tools have given Putnam an effective way to scale its marketing efforts in proportion to the explosive growth of the mutual funds industry. Putnam is optimistic that its shrewd use of CRM in conjunction with some predictive modeling software will help the company increase the lifetime value of existing customers by moving them into more coveted market segments. Putnam believes that its CRM initiative will also provide an increase in the number of new customers who buy classes of funds that are otherwise difficult to move in a saturated market.

⊙ IN SUMMARY

The CRM-related marketing tools available today create an opportunity for marketing departments to acquire and maintain a deeper understanding of their customers (through powerful business intelligence tools). These tools allow development of dynamic customer segments and profiles. But, this is only possible after identifying and prioritizing a department's CRM needs to match the overall CRM strategy. Then, and only then, can a marketing and customer interface be defined based on the unique characteristics of the company's markets, channels and products.

Innovative marketing executives can find new and exciting ways to integrate these powerful tools and tactics into their current marketing approaches. But the CRM team should take a people and technical inventory before investing in such sophisticated tools. Most marketing professionals don't have the experience or skillset needed to make the most of the analytical tools handed to them since there has been little precedent for them to need such skills. Therefore, most companies don't know how to make the best use of the customer knowledge they undoubtedly have within their data banks. Yet, even when the marketing professionals do have the skills to mine customer data storehouses, many times operational systems provide inadequate process and disconnects resulting in incomplete or misleading data.

Companies are realizing that something's got to give — customers will continue to demand faster and more personalized service, requiring more robust, better integrated systems. Ergo, a company can benefit from marketing automation tools, but first there needs to be in place a CRM strategy that envelops the entire corporate environment.

Chapter 11
Sales

As competition increases, customers become better informed, and products gain in complexity, sales professionals must grapple with new challenges. Sales-related CRM tools and sales force automation or SFA tools, can help.

Note: The sales and marketing departments must work hand-in-hand to deliver the ultimate customer experience, so the reader should read the preceding chapter along with this chapter.

Sales force automation tools, sales manager applications, product configuration engines and presentation tools are leading the march toward CRM. The business community views such products as a way to not only increase their sales force's productivity, but also contain and maybe even decrease the rising cost of sales. According to industry figures, the *average* cost of a direct sales call runs over $200 and continues to rise. As illustration: A late 2001, the monthly periodical, Sales and Marketing Management, conducted a survey of large, midsize and smaller companies and found that it costs an average of $189.75 to send a sales representative out on a sales call. In a February 2001 press release, Compaq Computer stated that cost analysis indicates that person-to-person selling can cost a vendor from $600 to $900 per sale.

General industry reports all show that after the introduction of a first class sales-related CRM system companies find at least:

* 10% increase in gross sales revenue per sales professional.
* 5% decrease in cost per sale mainly due to the ability to target specific suspects, prospects and/or customers (the "contact").
* 5% or more increase in closed deals because of the targeted approach allowing the sales staff to concentrate on contacts that are more likely to result in a "done deal."

Yet, the sales automation vendor space is splintered, making it difficult to choose from the many package options available, not to mention the challenges that await in putting these tools to work. To help the reader to better understand the sales automation product space, I've divided it into two basic groups. One works for the benefit of the individual sales professional by:

* Helping to prepare for sales calls by providing access to sales results and a contact's history.
* Providing dynamic and shared management of electronic agendas.
* Assisting in preparation of sales offers via product configurators and unique templates that aid in quote and proposal writing.
* Providing access to a knowledge base (usually maintained by the marketing or web department) to assist in the sales process.
* Formalizing call reports and gathering information into a central database.
* Entering expense claim forms and budget requests.

The other type of products assist sales management by providing:

* Detailed knowledge of the customer database and its segmentation.
* Objectives defined by segment, geographical area, seller, distribution channel.
* Real-time supervision of the sales force.
* Calendar consultation and organization of meetings.
* Follow-up on the processing of every sales opportunity.
* Detailed and/or consolidated sales analysis and reporting.

It may suit some businesses to limit a sales automation project to components that help the sales staff plan and follow-up on sales calls. But for others, taking a project to the next level (product configurators, proposal and contract generators, sales management modules, etc.) may be more cost-effective.

There are even sales automation applications that help sales professionals to not only *prepare* for a sales call, but to *make* the sales call; although many of these products require an investment in, and the use of, wireless technology and hand-held computers, laptops and PDA-like devices for mobile interaction. Throw the Internet into the mix and sales professionals can access pricing and inventory availability, prepare quotes in the field and even use closing tools to assist in preparation of the final contract. Never having to say "let me get back to you on that" definitely provides a competitive edge.

Put all of the sales automation products together and provide the means for them to work in tandem with other departments, especially marketing, and a business has within its corporate walls CRM's *piece de al resistance* — increased revenue per sales activity.

✪ PLANNING

As with all CRM projects, implementation of technology should come only after an automation plan is in place — planning is the key to a successful sales automation project. Begin the planning stage by taking an inventory to determine the sales department's current situation.

* Determine the marketing department's existing technology, resources and how it relates to the sales department.

* Identify everyone (within the sales department and company-wide) who has contact with prospects and/or customers: inside and outside sales divisions, remote and regional sales personnel, account managers, service and support personnel, secretaries and receptionists.

* Learn how each of these employees communicates with the prospect and/or customer. Is it face-to-face, telephone, email? What do they use to manage those contacts — PDA or computer-based address book, a typed or handwritten list of contacts, post-it notes, a contact manager, or what?

* Find out how each employee forwards information to the sales department. Is it via email, scribbled note, interoffice communication, voice mail...?

* What departmental path does customer correspondence follows? Is this correspondence archived where it can be easily accessed later by authorized personnel?

* Is there consistency and continuity to the correspondence? In other words, if an account manager has sent a customer a letter explaining billing discrepancies, does the sales representative have easy access to that letter? Is the same format used in all correspondence?

* How does the sales staff currently perform their duties? Analyze the sales process. For example, what do they do when they have a lead? How do they manage opportunities?

* Find out who knows when a sale is in the closing stages. If a contract is lost, who is informed and how? Does management determine why a contract was lost? If so, how?

* Note how other departments are kept abreast of when a sale is nearing the closing stage.

Only after sales management and the CRM team have a good grip on the existing programs and processes can the sales automation initiative move forward.

The Customer Life Cycle

The customer life cycle is a means of defining and communicating the way in which a company interacts with its customers and prospects, i.e. the acquire, retain and enhance continuum. The success of any company requires nurturing of this cycle and the sales department sets smack in the middle of the customer life cycle.

The marketing department provides the "acquire" impetus — the prospect — it's the sales department's job to ensure that the prospect becomes a customer. The sales and customer service departments, along with the entire corporate structure, then works to keep the customer satisfied enough to become a repeat customer (retain). Then, once there is an established customer relationship, the sales department, working in coordination with the marketing department, can begin to take advantage of the established customer relationship (enhance).

Where are the Breakdowns

Evaluate the current status of the sales department to determine where breakdowns might occur during the customer life cycle. For instance: What does the company's sales department need to fully participate in the customer life cycle? Can the sales department currently tailor its sales to meet customers' requests and to provide effective sales channels and staff to complete the deal? Check out how the department is currently handling customer relationships. The CRM team should follow the steps outlined in Chapters 2, and 8 (or pages 16, 17 and 89) to get a full customer's view.

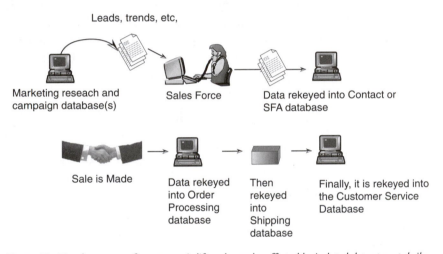

Figure 17. How key stages of a customer's life cycle can be affected by isolated departmental silos.

Understanding both the customer life cycle and the customer experience allows everyone to see the system breakdowns that currently occur. For example, the first breakdown usually occurs before the customer is "thrown over the wall" to the sales department. This is usually due to disconnects that occur between sales and marketing.

Marketing covers the activities required to get a qualified contact in front of a salesperson. Sales covers what happens from that point forward, thus the two departments must work hand-in-hand. Because these two departments and their functions are so closely related, it's only logical that the technology that enables them be integrated, i.e. highly leveraged sales and marketing processes working together, without any disconnects due to departmental silos, to deliver predictable revenue results.

The second breakdown in many sales departments is at the beginning of the customer life cycle, when the sales staff doesn't or can't respond quickly to an inquiry. The next breakdown is usually within the qualified prospects phase, when the quote is being prepared. Quotes require sales professionals to obtain accurate pricing, inventory levels and account information, which usually only reside in back-office systems.

The next area where a breakdown can occur is even more worrisome. It usually happens after the customer accepts the quote and places the order. If the proper systems aren't in place and aren't adequately integrated, processing the customer order will require a series of manual steps involving numerous departments, often resulting in duplicate entry, wasted effort and errors causing foul-ups, such as "I ordered five boxes, NOT five gross!" "What do you mean, you have no record of my order? I gave it to the sales person two days ago."

Document what happens and dissect the breakdowns. Use the results to drive the sales automation project.

Only after this evaluation process is complete can an automation plan be drawn up. But, before making changes in a sales organization, mark a path to follow. Be specific, don't give a stated goal of "we need increased sales." Articulate a clear objective or objectives. For example, "decrease low value leads by 20%," "reduce closing preparation from three days to one day," "increase forecasting accuracy by 25%" — you get the idea.

The automation plan should help the sales department to exploit the customer life cycle and obtain any other stated objectives. Then and only then can ensuing steps — technology selection, training, implementation and integration — begin.

✪ SALES AUTOMATION CHOICES

The business community and software vendors each realize that sales departments require automation tools to both access up-to-date information and to provide this information to others. If chosen with care, technology can help sales professionals to meet corporate objectives by aligning sales activities across multiple sales channels, allowing the generation of a single, global, up-to-the-minute view of sales and forecasts. With the Internet, mobile interaction can enter the picture — sales professionals can access quotes, compensation and real-time global sales forecasts from anywhere — for increased visibility into sales activities.

For many companies, SFA, which is among an elite band of fast growing CRM-related components, will fit their sales automation needs. A growing number of vendors offer a full complement of SFA tools (some in suite format) that can provide the ability to manage the sales process across many departments.

Note: Although many vendors offer products that purportedly fit under the CRM or SFA umbrella, some products are, in essence, only an upgrade of basic contact management software. Therefore, they may be only able to list key sales contacts, track interactions, and provide some level of performance measurement and contact analysis.

The many-module CRM suite products offered by some vendors provide the large- to enterprise-sized companies with the ability to implement flexible, customer-centric processes. Thus enhancing the overall corporate effectiveness by instituting efficiencies, and creating and nurturing long-term customer relationships. But, as stated in previous chapters, no one vendor offers a complete complement of products that meet the business community's needs as a whole; and many of the CRM suites don't have very deep sales automation applications. That being said, if there is a corporate-wide CRM initiative using a suite of CRM products, find a suite packages that provides sales modules which best meets not only the sales department's current needs, but also can grow with the department. Fill in any holes with third party products.

Sales Automation

The core of any true sales automation product is the interaction of a sales person with a prospect, turning that prospect into a customer and then maintaining a loyal relationship with the customer.

Whether the sales department's CRM initiative is based on a point solution, a complete sales automation package, or is part of a company-wide CRM initiative using the "suite" approach, the end results of any sales automation project must be the means to provide a single view of all sales activity. It must enable the coordination of all activities in the sales process, whether based in the office, field, or call center. True sales automation technology allows a company to bring together its sales force to interact in a cohesive manner, not as independent units.

The Communications Infrastructure

To work effectively, many of these tools need a robust corporate communications infrastructure that can support both traditional and wireless communication. With remote order entry and reporting, sales personnel can enter even the most complex orders directly into a corporate system enabling initiation of the order process immediately after the close. And with immediate access to the customer database, sales professionals can provide up-to-date status information on a customer's order at any point in the life cycle of a project or order.

The optimal solution will provide a web interface for data synchronization; thus, an employee working outside (or inside) the corporate environment can receive and transmit data, in real-time, using a standard browser. However, note that while real-time updates are vital for fast-moving industries (office supplies), in more complex environments, where the sales cycle is longer or more collaborative (turbine engines), a daily update may be sufficient.

Tools for the Sales Professional

The right product with the right customization is essential. Every sales department has different sales automation needs. (Just as some companies embrace a regimented, formulaic sales approach, and other companies ignore process and reward results.) Still, all have some common requirements, which should be met by every sales automation product under consideration. Look for products that provide quick and flexible configuration and customization so the sales staff isn't forced to *completely* change the way they work. A sales force automation package should provide easy and intuitive functionality (this is KEY).

Each of the following sales management tools should be assessed individually. Note that many of these tools will have overlapping or redundant capabilities, so choose carefully.

Account Management. Look for a system that gives a sales professional the tools needed to develop in-depth knowledge of his or her accounts with a (hopefully) comprehensive history of every interaction and event associated with that account. The product should also manage and track activities with potential customers and initiate new opportunities with existing customers. Find a product that provides quick call up of customer profiles, including all related contacts and deals, can drill down for a complete history of all activities associated with a particular contact, and can conduct real-time searches.

Contact Management. Choose a product that provides forms for creation of contact profiles, to-do lists, and appointment scheduling that includes alarm reminders. Require that the product offer the ability to have multiple contacts per account, insert notes to records, and embed documents (essential for most sales professionals). Also look for:

* Drill-down features to access detailed information on a particular contact and the ability to mark the importance of a contact via coloration, shading, icons, etc.
* The ability to view the relationship of a contact within the company and to view activity history by contact.
* Forms that can, for instance, create hierarchical charts to indicate business relationships and allow the user to outline the objectives, goals, and success factors for an account.
* Search functions that not only are intuitive, but also fast.
* Fluid movement between fields or screens.
* Configuration tools that can automatically assign and schedule activity plans and tasks including tailored lists with attachment capabilities.

Opportunity Management. Require opportunity management tools to be capable of assigning specific sales strategies. This might include:

* Automatically prompting sales staff to capture vital information about buying influences, competitors, advantage/risk factors and probabilities, and other definable factors.
* Coordination of sales strategies, events and activities to preempt competi-

tion by aligning the sales process stages and steps with customers' needs and timeframes.

* Empowerment of sales professionals to efficiently and effectively prioritize opportunities, spend more quality time with customers on quality opportunities, and transform them into revenue.
* The ability to provide reports on all current opportunities; listing their status, forecast probability of closing and total potential revenue.
* The ability to provide opportunity win/loss percentages as well as the reason why these opportunities were won or lost.
* Maybe even an opportunity data collector that could facilitate the entry of data through data entry forms and then upload to an opportunity database.

Time Management. Find a program that allows individual sales personnel to schedule appoints for any member of a sales team. With everyone moving in different directions, sometimes it's difficult to schedule meetings and appointments. However, with a group calendar it's possible to see everyone's schedule at a glance, eliminating double-booking conflicts. Many products even offer email notification of scheduled meetings and can link calendar events with accounts, contacts, and sales opportunities.

All products should provide:

* *Custom Reports.* A custom report generator that can provide reports in different formats — graphic, chart, spreadsheet, for example — is vital. Consider this a non-negotiable item, no matter what type of automation tool under consideration.
* *Email and Fax Management.* Observe how the product manages the broadcast of emails and faxes, and how it performs mail merge functions.
* *Infinite Fields.* Ask for an unlimited number of fields that can be easily defined by any user or, at a minimum, by a database administrator.
* *Library or Encyclopedia Feature.* Check document library capabilities, including synchronization and document sharing. Or integration capabilities, if this feature is only available from (say) within the marketing department's systems.
* *Linkage Capabilities.* Look for linkage or integration capabilities. Such as the ability to link with word processors, spreadsheets, accounting, email applications, etc. and to integrate with third party packages and marketing department processes.
* *Robust Security.* Require it to keep data safe from intruding miscreants.
* *Sales Team Management.* Enable team selling and management thereof by supervisor or team leader.
* *Other "Must" Features.* Don't forget easy customization of the look and feel of the interface and easy modification of pre-determined forms.

There are other features that might be a necessity for one sales department, but only a desirable function for another. Note that the following tools require more customization and development effort than the products in the previous list and are only relevant to specific selling situations. With that caution in mind, let's take a look at some of these tools and features:

Activity Management. This tool aids in the deployment of multiple, user-definable sales processes, including stages, events, activities, scripts, triggers, assignment and priority rules. Most products provide automated assignment, scheduling, escalation and notification of events and activities among sales team members. Look for ease of use — flexible selection criteria to create and view online or printed prioritized lists of leads, opportunities, accounts and activities.

Analysis Capabilities. Analysis engines are found more often in sales manager automation products, but individual sales professionals also use analytics to:

* Measure and improve the performance of campaigns — from lead generation to customer retention.
* Access valuable information, such as the quality of leads produced by various lead sources, competition by market segment, region, product and service line or item, and win/loss/deferral reasons at each sales cycle stage.
* Utilize graphical OLAP and sales management tools to help isolate potential market and customer base opportunities and buying patterns and monitor critical success factors.

Bulk Mail Management. Although this feature is found in many marketing modules, it is also useful in a sales automation suite. Many sales departments and individual sales professionals conduct bulk mail activities using both email and

Figure 18. OLAP can transform how a sales organization creates and distributes information for better decision making.

"snail" mail to generate opportunities. Look for a product that provides an opportunity database for further processing.

Closing Tools. Look for features that allow the sales professional to turn approved quotes/proposals into confirmed orders and contracts with a simple status change. Some closing tools offer the ability to generate a contract, create order entry forms and even submit the order to the appropriate people and/or department(s), etc.

Collaborative Tools. There are sub-categories under this designation. For instance, *Integrated Calendars* and *Event Management* are tools that enable collaborative, team-based selling along with providing sales personnel with a complete picture of sales and ordering activity. There are also tools that allow a sales department to involve the customers, for example, in product build and contract negotiations, so the sales professional can better accommodate their needs boosting the "closed sales" quotient. Collaboration is a boon for a company with a demand driven supply chain. These tools require integration to back-office accounting systems.

Configurators. This tool provides easy configuration of products and/or services based on specific customer needs.

Cross-sell and Up-Sell Management. This function addresses each stage of the customer's life cycle to make sure the company is achieving its goals both short-term and long-term. For instance customers' needs, preferences and attitudes that are captured through the Internet, mail, sales professionals or by telemarketing are input into a CRM database. That data is parsed using rules-based analytics and the results provide recommended solutions that are based on individual customer feedback, analysis, and strategic selling objectives. This enables the targeting of the company's customers with value-added services through cross-sell and up-sell opportunities. These recommendations, leads and follow-ups requests are managed through sales support systems. For instance, with a dynamic rule-based approach, the sales process can be tailored to an individual customer's financial profile, transactional history, learned behavior and personal interests. This personal involvement with the customers makes for regular communications and increased selling opportunities.

Customer Service and Relationship Management. This product provides the ability to see customer requests that are related to service and or maintenance and even other types of interactions. With such "heads up" the sales professional can manage his or her customers' queries. Some vendors even offer a tool that allows the sales professional to track each warranty activity, e.g. scheduled preventive maintenance, annual maintenance activities, and the like.

Expense Reporting. Concise expense reports are the bane of all sales departments, staff and management alike. With the right tools, however, it's possible to ease the burden while introducing order to the expense routine. Many expense reporting tools provide the ability to keep track of: advanced amounts, company and employee contribution and even mileage. Many systems even offer the ability to help management keep track of and manage all departmental expense reports.

Knowledge Management. These products usually offer integrated Internet, email and fax capabilities to automate distribution of products, services, pricing, promotions and competitor information. Most sales departments use this tool to keep prospects and customers informed about product and service upgrades, options, add-ons, replacements and substitutions to manage product life cycles, increase sales and reduce back orders. This is a great tool to leverage a corporate Intranet (or a website) to capture and deliver competitor information and other deal-closing knowledge to sales personnel, distributors and customers. Knowledge management software can also create web-enabled product and service catalogs, availability guides, and data sheets from an integrated database.

Lead Management. A function valued by many sales departments is the automation of the lead routing process, based on predefined business rules, whether to a sales team, channel partner or reseller. Lead management provides such features as guidelines, alarms and ticklers. Look for:

* Automatic qualification of leads against predefined criteria and prioritization via score or grade with colorization, shading or icons indicating level of priority.
* Tracking of leads via differing criteria (date, status, SIC code, campaign, product, territory, etc.).
* Reporting capabilities especially on results of lead tracking.
* Linkage features. Many leads come from the marketing department or a call center. Also the sales professional will also want to marry their lead management process with word processing, spreadsheets and the like.

Order Fulfillment and Service Management. These tools help track all activities involved in handling a customer's order until the products and/or services are delivered. They enable a sales professional to maintain the details of a customer's order and also the status of each order in terms of deliverables. Some products even offer a simple milestone/time-based project management and tracking tool.

Pipeline Management. Many sales departments need pipeline sales and order information made available to sales professionals and other qualified individuals. Comprehensive reporting by territory, region, product group, individual and line business should be a requirement. Many vendors also offer pipeline analysis, such as forecasting, sales cycle analysis, win or loss rates and other sales metrics.

Presentation Tools. These are products or functions that can be used to "wow" the prospect. Presentation tools are aimed at developing presentations and finding appropriate sales literature. Some products provide a presentation builder that helps a salesperson enter information, such as client, vertical industry, competitors, and products to create customized sales presentations based on reports and marketing literature published by other individuals within the company (marketing department and business development, for example).

Pricing Tools. These tools provide features that can apply flexible, user-definable pricing to address various needs by market segment, region, product line and service line. They can:

* Manage quantity breaks, customer-levels, contract pricing, limited-time offers, rebates, bill-backs and commissions.
* Automatically price quotes/proposals, orders and contracts to ensure accuracy and timeliness.
* Provide pricing and promotional information to employees, partners and customers via printed media, fax, email or over the Internet.

Quotes and Proposal Generation Management. This function draws upon boilerplate templates that permit sales professionals to easily create an impressive quote or proposal based on defined opportunities. Features should include:

* Easy generation of sales/pricing quotes and incorporation of said information into a proposal template.
* Ability to fax and email quote and proposal summaries with authorized signature designations and optional supporting details.
* Unlimited notes and document attachment capabilities along with easy, quick modification, recalculation.
* Ability to resubmit proposals to speed negotiations and approvals.

Remote Access. Most sales departments need a sales automation product that supports the efficiency and productivity of sales personnel from any location. By making available pertinent information from all necessary sources available in real-time, the process of information gathering, reporting and analysis is based on up-to-date, accurate data. This allows the sales professional to stay in the field AND in control of the information delivered to suspects, prospects and customers. For this to occur, though, there needs to be integrated web-based communications that allows access to a central data storehouse, along with tools to aid with analysis, forecasting, searching and so forth.

Report Capabilities. Every product will have some kind of report capability. But some advanced products can draw information from diverse data locations to produce, say, call reports, and/or sales versus forecasts, monthly sales figures comparison, profitability and even roll-up/drill-down reports.

Sales Forecasting. This software tracks summary forecast roll-ups of individual sales pipelines and account details, including status, advantage/risk factors, probabilities, and remaining steps to close. Many can also view sales forecasts and pipelines versus quotas by market segment, channel, region, entity, employee, customer, product or service line, and individual product or service item. Some products can even leverage sales forecast intelligence to improve product- and service-related resource requirement plans, corporate-wide.

Sophisticated Contact Management. These tools go beyond a run-of-the-mill contact management program to manage all types of business contacts, including employees, prospects, customers, and vendors. Such a tool might encompass handling the tasks of accessing complete online profiles of multi-entity, multi-site accounts, including contact relationships, activity schedules, account history and order entry. Another handy option allows inclusion of unlimited notes, attachments, and automated correspondence using integrated email or fax, as well as word processing and mail merge products.

Synchronization Capabilities. This includes data sharing via wireless devices. To avoid repetition, I refer the reader to the "Database Marketing" section of the previous chapter for more detail.

Web-based To Do List & Calendar. This feature has become increasingly popular with sales departments. It helps sales teams to schedule their activities and assists the team to maintain timelines for achieving commitments.

Automating the Sales Manager

Sales managers are responsible for up to 25% of a business's expenses and usually for more than 75% of its revenue. Yet, these executives, for the most part, employ only instinct and best guess to obtain results.

Over the last ten years or so, sales management has become more complex. The ratio of sales personnel to sales managers is on the rise while the average tenure of sales staff is on the decline. At the same time, sales departments are expected to increase revenue. In such an environment, the ability to forecast business accurately is crucial.

Software that can help sales managers and executives perform their duties more efficiently and effectively is one of the keys to a successful sales automation initiative. Technology helps sales managers deal with such complex functions as setting budgets, aligning territories and objectives, deploying salespeople, defining quotas and designing compensation plans, particularly in the large- to enterprise-sized companies with more complex sales organizations. There are also products being promoted by many vendors that can help sales managers to identify best practices, anticipate marketplace shifts, identify at-risk sales personnel, ensure commission programs are adequate, and more.

Some are just trumped up SFA tools, so it pays to be careful and do your homework. When looking at products that purport to offer sales manager features, look for the ability to:

* Manage individual opportunities and relationship information, such as probability metrics, amount of sale, closing date, competitors, etc.
* Manage assignment and reassignment of market segments, geographic territories, product/service lines and accounts.
* Automate distribution of leads to sales personnel based on user definable rules.

* Provide sales analysis (a full description of this feature is provided later in this chapter).
* Territory and account assignment and/or alignment.
* Coordinate multiple direct marketing and sales channels for both new customer sales and current customer cross-selling, including distributors.
* View individual sales staff reports.
* Generate comprehensive written and screen reports that provide information such as win/loss, close rates, probability analysis by individual staff members, etc.
* Generate sales metrics, such as lead source pipeline analysis or competitor analysis.
* Generation of analytical data on best practices of top sales personnel.
* Quota management (discussed in detail later in this section).
* Create and provide suggested guidelines for handling different sales situations.
* Benchmarking capabilities. Internal benchmarking tools allow a sales manager to see how the sales staff is doing against their peers. External benchmarking tools provide data on how the sales team is doing against sales department within the industry as a whole.
* Ability to provide roll-up sales by region, product, etc.
* Provide expense report management that include expense analysis and tracking capabilities by opportunity, campaign, special event, etc.
* Identify the right time to make the right offer to the right customer group through sales planning and forecasting.
* Provide compensation management capabilities. For instance, there are now tools available to reliably automate commissions accounting, while providing a full audit trail, financial modeling, and some even offer web-based reporting and management tools for incentive compensation plans.

Let's take a closer look at some of the more interesting features available in most sales manager software packages.

Incentive Compensation

The management of complex, perhaps personalized, incentive plans on top of shifting market demands and priorities requires a software solution that's flexible and at the same time can keep track of such items as goals and behavior alignment.

Incentive tools can design, create, maintain and administer incentive compensation plans and automate reporting tasks, allowing sales managers to proactively manage product line objectives, corporate and regional goals, territory considerations and dozens of other factors.

Incentive management software makes it easier to pay on sales over a certain target margin or other criteria (e.g. customer satisfaction, a click on a website, or the duration of a call in a call center), rather than just on revenues. With a incentive management component onboard, sales managers can perform model and what-if analyses on (say) the structure of sales teams, their roles and various compensation plans.

The right product can save a lot of time. One company found that on a monthly basis, it took seven to ten days with their old system to process sales commissions. Now, through the use of its new incentive management system, the company has reduced the process so it takes only two days to complete. The company also found it could perform the same tasks with one person, rather than the three that were necessary for the old system.

Quotas

Every sales manager is concerned with meeting sales quotas. There are now tools available to help the sales manager to set and modify sales quotas, analyze sales performance and improve the overall productivity of the sales department. Yet, to use this vast amount of data takes a complex set of tools.

Today's data flows from many directions — ERP systems, various CRM systems, field reports, point-of-sale, websites, third parties and so forth. While more data sources enable a finer grain of analysis, to use such tools is more difficult than the run-of-the-mill SFA program. But, if a sales manager possesses the necessary skillset and can conquer the complexity of these tools, he or she can drill down to levels as minute as a specific product number or as vast as looking at sales in a single geography or a whole group of products across several industries or channels.

With the right software and tools, a sales manager can define the characteristics of the department's overall sales environment, for example, the size of accounts, length of sales cycles, type of customers to be courted, and even how the sales staff is measured and paid.

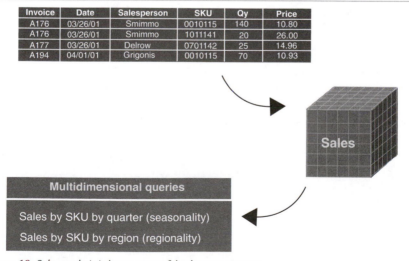

Invoice	Date	Salesperson	SKU	Qy	Price
A176	03/26/01	Smimmo	0010115	140	10.80
A176	03/26/01	Smimmo	1011141	20	26.00
A177	03/26/01	Delrow	0701142	25	14.96
A194	04/01/01	Grigonis	0010115	70	10.93

Sales

Multidimensional queries

Sales by SKU by quarter (seasonality)

Sales by SKU by region (regionality)

Figure 19. Sales analysis is key to successful sales management.

Based on the outcome, the right CRM tools can walk sales managers and executives through a specific process, ranging from quota setting via a numeric, regression model, or territory-based solution, to a more large account opportunity-based quota setting solution. The manager or executive can also explore what-if scenarios, analyze them and easily modify quotas.

Sales Analysis

Sales analysis tools provide some of the most important and useful information available to a sales executive and/or manager. By tracking, analyzing, and collecting sales information, sales managers and executives can discover business opportunities and possible weaknesses within not only the sales department, customer service department, field service, etc., but also the products and services being offered by the company. Sales analysis can also uncover potential areas of growth.

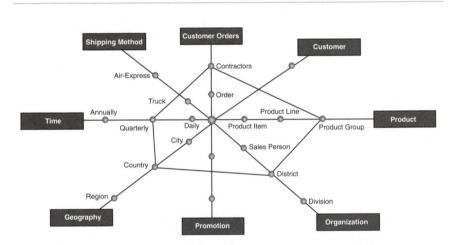

Figure 20. Sales analysis tools can provide data on any number of selected categories in conjunction with specified criteria, such as period of time, geographical location, promotional effort, product and so forth.

Territory Alignment

For a company with a complex sales organization, new sales channels and/or mirrored sales structures, tools that aid in territory alignment are a godsend. These tools can help with deployment of sales staff and aid in building a structure and foundation for success. There's no limit to the information that can be included. Most tools offer:

* Profile information.
* Sales information.
* Market sales data.
* Information on market potential.
* Information about call history.

* Specific information about whether a certain account is a target.
* Help in determining the value of different customers or customer groups.
* Management of multiple sales teams within the same system.
* Deployment of multiple map windows so a sales manager or executive can visually compare one team's alignment to another and see how they might need to coordinate.

Reports

The many reports that can be found in most technology-assisted management products might include sales force overlap, scenario disruption overlap, comparison, scenario graph and layout. Some tools even have an index that measures different types of items with different magnitudes. In the health care field this could consist of physicians, prescriptions, managed care information, hospitals and accounts, and make them all relative to one another as components. The sales executive can then decide which ones to stress.

○ VERTICAL SOLUTIONS

Many of the *efficiency* challenges facing sales departments can be addressed by general purpose sales automation tools, but these products usually don't functionally address the *effectiveness* challenge that the modern sales department deals with every day. The CRM and sales automation vendor communities have listened and learned. Today, it's possible to find products designed to help the average sales professional working within a specific vertical industry with such tasks as needs analysis, product configuration, ROI justification and proposal generation that reflect their unique industry requirements.

If a sales department's job is to sell telecommunications equipment, it has an entirely different sales process than a sales division that seeks customers for pharmaceutical goods. In the same vein, a product designed to provide the systems and processes to enhance the productivity of a person selling perishable goods won't work for a company that sells insurance. (See the "Vertical" section of Chapter 16 for more detail.)

○ THE HUMAN ELEMENT

Don't begin a complex, expensive sales automation project without first consulting the sales staff. After all, they're the ones who will be using the technology. Without the end users' input on selecting and implementing these tools, companies can find their sales automation initiative stalled.

With so many different sales automation systems available, the CRM team should be able to quickly find a product that can be easily customized to meet the unique needs of the sales department. When implemented correctly (i.e. comprehensive training and astute customization) the right components can no doubt improve sales. Yet, the hopes that initiated the enormous investment in the latest technology to follow sales leads, acquire new customers and boost the bottom line can be quickly dashed, if the "people factor" isn't given enough support. Re-read Chapter 6 for a thorough discussion on this subject.

Achieving a smooth transition means listening to and addressing the various needs of an often reluctant sales force. Planning for the human element is even more critical when introducing technology into a sales department. Numerous business-es (large and small) have found that the promise of technology-assisted sales is unre-alized largely due to failure to factor in the end users — the individual sales person.

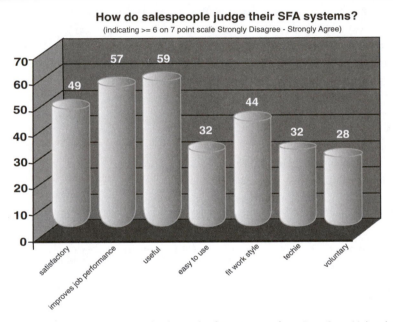

Figure 21. Courtesy of The Institute for the Study of Business Markets, Penn State University.

Motivate Don't Mandate

Motivate the staff, bring them into the decision making process, provide adequate training and continuous hands-on support. Also provide tangible incentives to help motivate the staff to accept the new automated systems. Institute rewards for use of automation tools. Educate the staff on the positive business results, increased sales and better compensation that can result from the use of the new systems and processes.

In my experience the motivational approach works best because sales person-nel are, by their very nature, independent gregarious individuals who make a liv-ing out of finding ways to work around perceived obstacles.

A mandated approach usually doesn't work with most sales staff. A mandat-ed change in a working environment means the user either accepts the new way of doing things or is forced to leave. When a sales force is forced to learn a new system and enter in volumes of data on their customers and prospects; they will rebel in one way or another.

All Employee Aren't Equal

The key error many sales executives and CRM teams make is viewing all sales professionals as being equal when it comes to the use of technology-assisted sales systems. There are actually three different basic employee types, each with their own set of reasons and motivations for using a new computer system.

The first type is the indentured sales person, such as the typical call center agent. This group has no choice, they must use whatever system that's provided to them, no matter how inefficient, difficult or confusing it may be to use. If they can't learn to use and work with the system provided, they won't have a job. (This is the one situation where a mandated approach will most likely work just fine, although a motivational approach might provide a better success rate.)

The second type is the department's computer "geeks." This type views computer systems as an opportunity to learn something new and are usually eager to take on a new challenge. Most are "closet programmers" to a certain degree. Often, this is the group the CRM team turns to for user feedback on proposed and/or new systems — a bad approach — this group doesn't represent the typical sales professional.

The third type are the volunteers. This group already excels in performing their duties without the help of CRM or technology-assisted sales tools. Even with outdated reports, post-it notes and snail mail, they succeed in getting the job done. Members of this group still manage to regularly launch new products, find new prospects and close deals the old fashion way. This is the group that should make up the bulk of any test group. To get the volunteer to adopt a new CRM tool, he or she must be convinced to *volunteer* to do the job in a different way.

✪ BENEFITS

There are numerous benefits of CRM in a sales department, some of which have been mentioned previously. Let's take a closer look. For example, some tangible benefits that accrue with the implementation of a CRM project within a sales environment are an increase in:

* The amount of time spent with existing customers (which can be measured by the number of daily calls or hours spent with existing customers).
* The number of suspects and prospects pursued (which can be determined by measuring the number of new contacts versus existing contacts by each sales professional on a per day, week, month, quarter or annual basis).
* Responsiveness metrics when it comes to requests received within a sales department whether from prospects, customers or internal sources. E.g., information

turn-around and follow-up (which can be measured by taking the number of days between initial contact and the date follow-up occurred whether by sending requested information, telephone contact or personal visit).

✳ Revenue per month (difficult to measure since management must ensure that the time saved through automation is used in a productive manner. It will also vary from one salesperson to another).

✳ Close rates and at the same time a reduction in the time it takes to bring about the close. (Again, difficult to measure.)

There are also intangible benefits, which should be considered when looking at implementing a CRM solution for a sales department. For example: CRM systems improve communication with the customer and, if fully integrated, within the company. An integrated CRM system can provide quicker access to necessary data — data that's fresh and topical.

Now let's take it a step further and put some numbers to these benefits.

✳ Senior financial advisors at North Shore Credit Union reported in 2001 that through the use of a Pivotal solution suite, NSCU has increased its sales revenue in fourth quarter 2000 by almost 20% over the previous quarter.

✳ Independent surveys of Invensys CRM users report 10% to 15% reductions in sales cycle length, 30% reductions in account turnover times, and up to 20% gains in sales revenue.

✳ An Insight Technology Group report shows that a properly executed sales-related CRM initiative can provide increase sales revenue by up to 40%, cut sales cycles by 30%, improve margins by 2.5%, and decrease cost of sales by 20%.

✳ Knight Frank proudly announced in 2001 that it has increased its sales revenue and profitability while enhancing the quality of service its commercial property customers receive. This was after the company standardized on Siebel applications within its commercial property division.

✪ IN SUMMARY

In today's rapidly changing world, many businesses are struggling with how to rationalize their sales resources and maximize the return on investment (ROI) in finding, acquiring, and retaining customers. At the same time, sales productivity is being severely scrutinized leading to emerging technology-based solutions that combine the best of "old and new" approaches to achieve high levels of sales productivity.

The true objective of any sales automation project is to create more time for sales professionals to sell. But, in reality, at the end of the day, it's really all about the bottom line, and that's what a well-planned sales automation project delivers — sales results — profits — satisfied stockholders.

Chapter 12
The Call Center Evolution

D espite all dire predictions, the reports of the death of the call center are greatly exaggerated (to paraphrase Mark Twain). The new century brought with it a more competitive environment — new technologies, a spat of deregulation and the precipitous rise of the Internet. With so many businesses facing stiff competition, the demand on the corporate call center (the primary touch point for the vast majority of customers) is skyrocketing. What this means is that the call center must be able to deal with this new paradigm. It must be able to handle more calls, which are often more difficult and time consuming. This requires hiring and training a call center staff that is more knowledgeable than in the past and then providing that staff with access to more information.

❂ THE STAMPEDE TO CRM
Businesses must be prepared to face stronger competition, higher customer expectations, cost-containment pressures, and broader and more complicated product and service offerings.

As illustration: Insurance companies now have financial institutions offering products and services that were once their domain. Banks face new competition from insurance companies, mutual funds and mortgage companies. The telecommunication industry is reeling from deregulation and the influx of wireless providers, cable companies, and the like into their space.

Let's not forget the affect of the Internet on the entire business community, bringing competition from previously unknown sources — new companies (Amazon) and remote businesses (Home Depot, through its website, can easily compete with the local hardware store).

Due to the new competitive climate, businesses find themselves forced to sell a wider variety of products and services, with many of these offerings more complex than in the past (remember my example of customer-centric product configurations related in Chapter 8). To be competitive, poor (?) AT&T and its offspring now must offer not only local and long distance service, but also Internet access, DSL service, wireless communications, etc. Gateway Computers can't just offer ten different computer models with set configurations; instead, it must adapt each product to meet an individual customer's need.

The unrelenting need to generate more revenue in the face of a fierce competitive environment has caused a stampede toward CRM initiatives within the call center environment. Since CRM enables any call center (inbound, outbound, or blended) to instigate targeted cross-sell and up-sell opportunities even a "cost center" can become a "revenue center." At the same time, the call center can up the level of customer service provided to the company's customer base. In other words, implementation of CRM within the call center environment is a win-win situation.

How does it work? By using a CRM strategy that works in partnership with traditional call center processes and technologies. For instance, with CRM working in partnership with IVR and ACD systems, the amount of time agents spend between telephone calls can be reduced by 10-30 seconds; resulting in more available agents and less time a customer must spend in queue or listening to a busy signal. That might seem insignificant, but over an 8 hour shift, that 10-30 seconds makes an agent available to handle an additional 10 calls and in a 100 seat center, that's an additional 1000 calls every shift.

There is also the recent trend toward a blended center model. In a blended center, a manager can quickly increase the number of agents during peak calling time or unexpected busy periods. This is easily accomplished by shifting agents from outbound calls to incoming calls, or from email or chat duty to inbound call duty, or by utilizing an agent working in a remote center or from home.

The Bounce Syndrome

Traditionally, call centers have been on the fringe of the corporate environment; thus, their systems were rarely sufficiently integrated with other corporate systems. But this lack of integration results in what is euphemistically referred to as the "bounce syndrome," which refers to the act of transferring a customer from one agent or department to another in an attempt to find a solution for the customer. Although the numerous call transfers generally result in a solution, it typically leaves the customer frustrated due to the circuitous and lengthy route taken to arrive at the solution. Integrated CRM systems reduce the bounce syndrome. With the right systems in place, a customer can be identified based on the source of the call, the relevant customer and product information can be retrieved and the call can be routed to the most appropriate agent or department.

Of course, the call center's CRM system must be fully integrated into the corporate systems. That's the only way to provide seamless customer service. Here's how an integrated CRM system can eliminate the bounce syndrome, increase sales, customer satisfaction, and the overall corporate profit margin:

The system identifies an incoming caller as an IT supervisor from a small, but valuable, business customer that recently ordered 20 identically configured computers. (How did the system know this? By querying integrated corporate-wide data storehouses to garner available information about the caller.) It immediately routes the call (using skills-based routing techniques) to an agent with specific knowledge about that product's usage within a business environment.

Let's take this scenario a bit further. Say the caller wants to return one computer. With the right system in place, the call can be transferred to the appropriate person/department along with all relevant data — account number, contact information and the reason for the return.

But, now assume that this same customer had sent an email inquiry the day before about a new computer that had just hit the market. With a fully integrated system, the call center agent could be and should be aware of this inquiry. For this same agent to put on a salesperson's hat, he or she needs access to a knowledge base that can provide timely and accurate information about not only the product's specifications, but also marketing campaigns, specials, and so on.

Next, suppose that the customer responds enthusiastically to the new product after listening to the agent's spiel and wants to purchase three. The agent can now access inventory to see if the product is available, and the shipping department to advise the customer when the product will be delivered. What began as a customer service call became a sales event — integrated technology made it possible.

However, perhaps the sale shouldn't have been made at all because that customer's company had a large number of outstanding invoices and had been placed on a "credit hold." (Even when the corporate-wide CRM systems are integrated, if there isn't an *overall*, corporate-wide systems integration, there can still be disconnects.)

"At Risk" Customers

In another situation, a customer sends a fax to accounts receivable on Monday morning responding to a known billing problem and that afternoon receives a call from the collections department demanding payment of that same bill. What results is an angry customer and an embarrassed and apologetic collections agent (read "call center agent"); all because collections (which is usually a call center) didn't have real-time communications with the account receivable department. This type of situation can quickly place a customer in the "at risk" category.

Since I've neatly segued into the "at risk" category, let's consider how a CRM system can help to identify an "at risk" customer and turn the situation around.

First, upon receipt of a customer contact, a CRM system can immediately identi-fy the customer, collect and analyze data, and indicate whether the customer is at risk or not.

The system separates any at risk customers from the pack. In one instance, it might bypass queues and IVRs to provide the customer with special, hassle free service. In another instance it may pro-actively place an outbound call to extend a special promotion designed to retain the customer. Yet, at another time it may for-ward the customer's data to the sales department for face-to-face interaction with the customer. Without integration of the call center's systems with company-wide systems, none of this could occur.

Improving the Customer Experience

Never forget that every contact between a company and its customer is a chance for the business to learn more about the customer and to deepen the relationship between the two. At the heart of any CRM initiative is the need to improve the overall customer experience. CRM, working with traditional call center technolo-gy (ACDs, ANI, IVRs, etc.), is the answer. It can better manage a customer interface by, for instance, enhancing call routing and tracking. It also can provide a single repository of information, negating the need for customers to repeat themselves. Any employee (including call center agents), in any location, can quickly and effi-ciently address a customer's needs through information provided by an integrat-ed CRM system.

Call centers deal with ongoing customer activity (where is the order), answer questions (how to cook a turkey, remove a stain, fit a square peg into a round hole), participate in telemarketing campaigns (surveys), initiates telesales, resolve issues (accounting problems, product defects and bugs, delivery questions, service issues) and collect outstanding bills. Whether the customer contact is via landline telephone, an IP-based call, website chat, email or fax, they're demanding high-quality interactions and personalized levels of service and support. This particu-larly holds true for the most affluent customer contingent — the ones the busi-ness community most wants to attract and retain. This favored segment demands instant access to information and service via whatever channel they use — web (FAQs, searchable knowledge bases, call-back services, voice and text chat), email, landline telephone, wireless, etc. — and they want consistency across these chan-nels. This customer group is also very vocal in their desire for personal service — they demand that the companies they do business with know who they are and understand their needs.

To provide this level of service, a real-time, 360-degree view of the customer is necessary notwithstanding the channel of communication. To accomplish this goal, businesses are designing and implementing CRM strategies and technolo-gies in the call center environment that can:

* Provide consistently high-quality service regardless of channel.
* Feed quality data to a host of corporate departments.
* Provide real-time information, customization and personalization to customers.
* Use a robust customer-centric information database.
* Integrate with existing front- and back-office legacy systems.

✪ CRM AND TRADITIONAL CALL CENTER ARCHITECTURE

In the past, astute executives implemented call centers with a complement of staff, business process and enabling technology. Yet, most call centers were considered to be a low-level business function with minimal importance, so the center was built to provide adequate customer service at the lowest cost possible.

The traditional call center, built to handle voice calls only, is evolving into what some pundits have designated as a "contact center" or "multi-media center," which can handle customer inquiries through any contact device — letter, telephone, fax, email, Internet, wireless (PDAs, WAP telephones, handheld computing devices) and perhaps even television. The quandary for call center managers is how to go about prioritizing contacts via these diverse channels. For instance, which gets priority? A landline call from a new customer (which must be answered immediately)? A WAP-based instant message from an established customer (which normally waits a minute or two)? Or an email inquiry from a valuable customer and stockholder (which usually waits upward to 24 hours before being answered)?

It's not difficult to see the dilemma most call center managers face. Include live chat sessions, collaborative web browsing via a high speed Internet connection, and factor in the pressure for inbound call centers to become profit centers where a service interaction becomes a sales interaction, and anyone can see the nightmare situations the typical call center manager must deal with daily.

The Time is Right

Now is the time to start thinking seriously about enhancing call center operations. CRM requires companies to invest in specialized software to record, track and analyze customer interactions. For instance, databases are needed to store information along with data modeling and profiling tools to not only understand transactions, but also to predict buying patterns and to launch customized offers during any subsequent customer contact.

CRM uses sophisticated voice and on-line media routing hardware and software to direct customers to a person or to an interactive voice response (IVR) system or web-based self-service information to get answers to questions, fulfill orders, take comments, etc. CRM tools leverage a customer's identity, along with the nature of the contact and any special attributes (such as the customer's spoken language) to intelligently direct queries.

CRM vendors are eager to sell products and the business community is an eager buyer. Since 70% to 80% of call center costs are people-related, many corporate boards are willing to sacrifice short-term profits for new investments that will deliver savings in the long-term — enhancing a call center is the best way to reduce costs while improving the quality of the customer relationship. Investing now, rather than later, can give a business a competitive edge.

Metrics Win the Day

Most call centers measure success based on productivity metrics. While such metrics aren't the best measure of call center effectiveness, they offer the hard numbers needed to get an investment in a CRM initiative approved. Investments that result in real dollar savings and/or cost avoidance are the only ones an executive board will readily accept. Acceptable investment justification metrics, for example, can include:

* ✳ Reduction in staff through productivity improvements.
* ✳ Call and email deflection and/or reduction.
* ✳ Reduction in training and hiring costs.
* ✳ Real estate expense savings (less staff, less space needed).
* ✳ Telecom expense reduction.
* ✳ Increase in customer revenue.

This is just a sampling. There are many more, some specific to individual call center environments.

Average Call Center Costs	Average	Range
Phone call:	$5.50	$2.00 to $12.00
IVR:	$0.45	$ 0.25 to $ 1.00
Email:	$5.00	$2.50 to $40.00
Web Self-Help:	$0.24	$0.05 to $ 0.50
Web Collaboration/Chat:	$7.00	$1.50 to $12.00

Figure 22. This table summarizes the justification criteria acceptable for different types of call center activities. Source: Call Center Magazine.

❂ HASTE MAKES WASTE

The objectives of most call center investments are to reduce costs, improve quality, and increase productivity and/or revenue. Successful call center investments must simultaneously address people, process, and technology. Introducing a new technology or application, without analyzing its impact, can doom the effort to failure. Even in the best of times, call centers cannot afford mistakes; in a challenged economy mistakes are unforgivable.

To be successful, the modern call center must be able to intercept all channels of communication (telephone, chat, email, regular mail, fax). At the same time it needs to be able to provide inbound, outbound and blended capabilities.

Traditional call center technology can handle some of this; but to evolve into a hub for customer contact, the call center also needs an integrated CRM system. CRM integrates all customer interactions and data, no matter what the channel, providing a single view of each customer relationship. Today that's not happening. (A leading independent search firm, Forrester Research, did a survey, which found that more than 80% of call centers have no idea if a customer has touched the company's website before calling in.)

As call centers evolve into multichannel contact centers that handle both telephone and on-line communication, investments in tactical online or e-service solutions (email response management systems, web self-help, and bots) becomes a necessity. Such solutions, when properly implemented, can improve quality and productivity. However, too often, these projects fail because companies do not carefully analyze needs, resulting in the wrong tools being implemented.

CRM initiatives also can disappoint when companies take short cuts, skip testing during implementation, and do not garner adequate knowledge and resources. These failures create incremental costs instead of savings. For example, web self-help products (which can pay for themselves in a couple of months) might be a great way to relieve the burden on a call center. But, there's a catch — the success of web self-help products depends on having a pre-defined database or library of answers or knowledge. To have the correct answers, the company must accurately anticipate the questions its customers will ask.

Many self-help implementations fail because the business didn't determine its customers' needs before building the application. Some companies make the classic mistake of answering questions they *want* their customers to ask, instead of responding to what customers *actually* ask. It's a costly and sometimes fatal error. Customers confronted with worthless information will abandon a site and go to a competitor. Or, they'll revert to calling the call center, at a cost to the company of at least twenty times that of a web self-help transaction. Minimize the CRM risks through careful planning prior to implementation of the applications.

A "Contact" Center?

Okay, I give industry gurus their due — today's call centers are evolving into "contact centers." However, most industry insiders continue to use the term "call center" (somewhat akin to the aphorism, "dialing a number," when in reality the number is probably "punched," rather than "dialed").

⊙ EVALUATING CRM SYSTEMS

The real power of a CRM-enabled call center is that each customer issue can be matched to the agent who is best able to service that customer. Admittedly, much of this is accomplished via traditional call center procedure wherein it's possible to identify and pull a customer's records by matching information such as a contact identifier — an in-bound 800 number, email address or customer telephone number. CRM fine-tunes the process, which may be as simple as connecting a customer to an agent trained in the product or service that is at issue, or as complex as taking advantage of an agent's education, people skills, and experience to deal with an escalated problem. Most importantly, high-value customers can be identified immediately and receive high-priority service.

Look for vendors that provide not only full call center functionality including IVR and CTI, but also connectivity to PBX/switches, VoIP, email management, self-service, media blending, skills-based and rules-based routing, interactive text chat, assisted browsing, and workforce management.

While features and functions are important when evaluating various CRM systems for installation in the call center environment, just as significant is the ability to implement the product(s) quickly. The quicker a system is up and running, the sooner superior customer service, higher revenues per customer, more retained customers, and better agent productivity rates can be the norm.

In evaluating a call center-based CRM system, the ability to provide the following attributes should be considered non-negotiable:

* Creation of and access to customer self-service features on any company website.
* Access to interactive customer support elements which enable communication between a customer and a call center agent (i.e. chat, call-back).
* A robust email management response system.
* Management of all forms of customer communication.
* Queue capabilities — incoming customer inquiries, as well as scheduling and assigning agents by various criteria.
* Recording of incident data and automatic assignment of any incident to the appropriate person or department.
* A view of each customer's service history.
* Access to information about problem resolution.
* Automatic inquiry assignment.
* Ability to manage problems that arise during the inquiry resolution stage.
* Automatic escalation of an inquiry, when needed (based on user-defined criteria).
* Appropriate data recording capabilities and management including tracking and monitoring.
* Database management including statistical analysis and reporting abilities.
* Ability to integrate with existing technology, processes and systems.

A Blended Center
A critical component of a 21st Century call center is the universal queue. In the

optimal call center, this central system receives an inbound contact, regardless of channel (telephone, email, fax, web chat, or call-back request), and routes it to the appropriate call center agent based on established business rules. Along with the contact notification, the agent receives a 360-degree view of the customer's contact and transaction history as well as a recommendation for dealing with the inquiry and identification of sales opportunities.

A successful call center's CRM system will have the ability to effectively integrate inbound (customer initiated contact) and outbound (company initiated contact) communications. Rather than building and maintaining two separate call centers (one outbound, one inbound), with the right technology in place, both outbound and inbound communications can be managed in an integrated fashion. This is sometimes referred to as a "blended center".

Telemarketing and/or Telesales

As more and more inbound call centers make the move toward a blended center, they must morph their functions to handle outbound calls as well. This requires traditional technology working with CRM components to handle typical functions that will enable the business to:

* Create a call list and assign calls to an individual or group responsible for completing outgoing calls.
* Access dynamic, branch scripting and scoring based upon script responses.
* Perform auto-dialing.
* Record call information.
* Plan calls, supported by access to pertinent information via attached documents or on-screen information.
* Monitor call activity through call statistics and call reports.
* In addition to order taking capabilities, track customers' orders so as to provide a speedy response to customer inquiries.

Sales Configurators

Identifying the right products and services for agents to sell to a specific customer presents another challenge for call center management. The ability for an agent to access customer and account information while a customer is on the telephone is helpful in this regard. But, a growing number of call centers are also considering the use of a typical CRM component — a sales configurator — to help automate the sales process. Sales configurator applications run atop a rules-based engine, which helps to translate a customer's anticipated desires into a sales opportunity and, at the same time, simplifies the selling of complex products and services. With sales configurator technology, companies can rapidly roll out new campaigns and products without in-depth training for call center agents. In addition, inexperienced agents and even new hires can intelligently present very complex products and services.

Interactive Voice Response (IVR)

IVR technology has been in use in call centers for a number of years. It's not strictly within the domain of CRM, but rather works in tandem with CRM components. It enables customers to respond to voice prompt menus using a touch-tone phone or verbal input.

Simply put, an IVR is an automated voice system that responds to telephone calls and offers callers keypad or voice driven options. These menus typically lead customers through a number of different processes, including placing orders, providing verification information, making inquiries and self service ("1" for checking, "2" for savings, "3" for credit card account). The IVR can then route and prioritize calls based on the caller's input, which can help to reduce the duration of a call. Note, however, that unless the IVR is programmed intelligently, the process can be cumbersome and intrusive.

Cross- and Up-Selling

A 2001 NelsonHall (an IT services and business process outsourcing tracking company) report predicts that companies will focus on improving customer enhancement capabilities over the next couple of years. One of the weakest link in any call center-based CRM solution is the ability to integrate customer data across multiple channels so it can be used at point-of-sale to cross- and up-sell products and services.

A call center-based CRM system should proactively manage customer interactions to differentiate service, and enable cross-selling and up-selling, regardless of channel. It should also be able to integrate the web with the call center and its systems, giving the call center agents instant access to personalized customer information they can use to cross-sell and up-sell products and services consistently across all channels.

With a CRM solution that provides a dynamic rule-based approach cross- and up-selling can be tailored to the customer's financial profile, transactional history, learned behavior and personal interests. Personalizing the customer experience helps foster customer loyalty and therefore customer retention.

Of course, all of this requires a CRM system that supports sophisticated customer information handling and customer analytics within the call center environment. So, even call center-based CRM systems must support customer analytics. This means that customer analytics must move out of the back-office (where the only concern was product development) to a front-office discipline providing real-time advice for handling customer interaction regardless of touch point or channel. This is what really enables the most advantageous up-selling and cross-selling opportunities.

Data Management, Reporting, and Analysis

A call center is usually the primary point of communication between a business and its customers. Its systems and processes must not only effectively capture and report all relevant information (each telephone call, email, chat session), translate the data into meaningful reports and feed it to the CRM system (e.g. intensive database requirements), but it must be closely integrated into the overall corporate CRM systems.

Another CRM function is to capture customer feedback. Much of this feedback will consist of complaints; but, as a call center sage once put it, "the complaining customer is a company's best friend." The optimal CRM system should help a company to use customer complaints to build consistent resolutions for grievances.

The CRM system should also enable management to determine, for example, what products fly off the shelves and why (price, appeal to a specific customer segment, no competition, or what?). Conversely, it should determine why a specific product doesn't sell (ineffective marketing, wrong target audience, pricing, competition or ??). The information doesn't have to be limited to products; it should also provide insight into, say, a marketing campaign — documenting its success or failure and helping to understand the "whys" of both.

There is so much data collected by call center systems that a coherent storage method, such as attaching a record number to each record is a necessity. Response(s) should have the same record number.

The data feed shouldn't go only one way. The call center must also be able to receive data from other sources (marketing, sales, account, or research and development) so the call center agents can always deliver the right information to the customer.

For such a vast amount of data to be of any use, reporting and analysis techniques are needed. They play an important role in call center management.

Call center analysis grows in importance as operations move from basic measurement techniques (i.e. number of calls answered, call abandonment rate, call duration, etc.) to more complex models that take into account such details as:

* Up-selling and cross-selling.
* Customer satisfaction.
* How many suspects, prospects and customers called the call center?
* Details about who called.
* The products or campaign that triggered the call.
* The type of problems encountered and the specific solutions offered.
* Operational data (average response time, abandonment rate, elevation/transfer rate and call duration), which is vital for forecasting and understanding performance metrics.

Data management, reporting and analysis requires strategies, processes and technology that enable a call center to not only automatically log every contact event and agent response, but also allows agents to add notes on customer issues and the resolution for every transaction.

Analysis Tools

OLAP systems store and access data as dimensions that represent business factors like time, functional activities, and geographical location. This information is stored multidimensionally, i.e. like a cube that can be viewed, turned, and shifted from any angle. Because information is presented in a business context rather than a database context, it is more intuitive. OLAP reports that take an analyze-then-query approach allow decision makers to access data the same way they identify and solve problems — by reviewing totals or summary information first, then looking at the underlying details.

Some business might benefit from *multidimensional analysis* (aka multidimensional on-line analytical processing). If so, look for an application that provides a special tool for building cubes that provide multidimensional views of data, allowing all information to be arranged around a specific element or set of parameters.

Another useful tool is an *interaction analyzer*, which is used to gather business intelligence through standard report and ad hoc analysis. Customer interaction analysis and reporting include, for example, agent activity and reporting, and campaign results analysis and reporting.

Trend analysis is vital and many CRM solutions enable users to gather and analyze data for use in product development, product design, sales and marketing. Database analysis capability should also include segmentation and profiling, cluster analysis, market basket analysis, and predictive modeling and scoring using

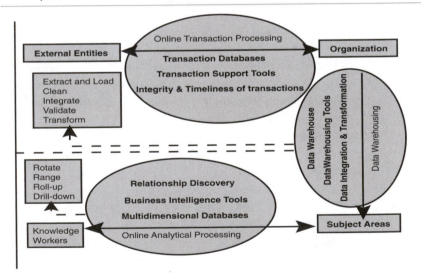

Figure 23. Multidimensional Analysis. Multidimensional databases store data in a multidimensional structure on which analytical operations are performed. Queries posed are quite complex and require different views of data. The queries may either be answered from a materialized cube in the data cube (see Figure 15 in Chapter 10) or calculated on the fly.

both customer data and input from the web and the call center environment.

The ability to carry out sophisticated database analysis to provide concrete recommendations to enhance corporate decision-making is essential to the success of a CRM system. The data gathered by CRM-enabled call centers is golden and companies can use that data to look back, analyze trends in customer issues and identify problems (such as whether the issue or problem is related to quality, product development, customer service or accounting).

Reporting Tools

Detailed reporting capabilities help a business monitor the quality and responsiveness of its customer service efforts providing insight into a company's overall operations. Information such as number of interactions or incidents by product can be used in product development efforts.

Nearly all call center solutions provide some reporting functionality, but CRM solutions can optimize the use of the data gathered by call center systems. For example, traditional call center systems don't provide real-time reporting. Many CRM products do — managers can have up-to-the-second detail on resource allocation and activity queuing. Check to be sure the reports the CRM system offers are dynamic and instantly viewable on the call center manager's and/or supervisor's computer screen, NOT a repaint every 10 seconds or so.

Report writing tools should:

* Be graphics-based with drag-and-drop functionality.
* Focus on customer acquisition, behavior patterns and segmentation; conversion and retention metrics; promotional effectiveness (ad evaluation, special

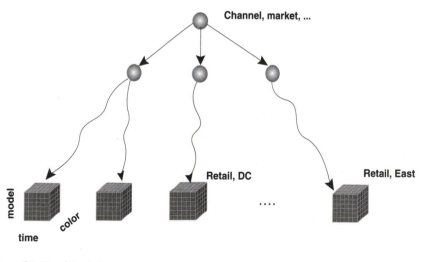

Figure 24. Trend Analysis.

pricing, etc.); and perhaps merchandise assortment.

* Allow the users to navigate and search high-resolution, personalized content in reports that are viewable in a pure browser environment without the need for plug-ins.
* Enable programmers/developers to build reports that access any data source, do any calculation, compose any format, and seamlessly integrate with any system.
* Enable management to generate reports on demand (to be viewed immediately). Although in some instances, generating a report on a schedule and caching it (to be viewed later) can be acceptable.

No matter what data management system used, it should be XML compatible. This will enable applications to incorporate data functionality from third-party websites and applications.

Computer Telephony Integration (CTI)

Some businesses find that a vital ingredient for their CRM initiative is the ability to pool updated customer information with knowledge gained from a community of other individuals (who share similar tastes and interests) in order to provide predictions of future customer behavior. Of course, such a solution also should be able to display a customer's profile, contact history and any other pertinent documentation. The problem is that most of this data is located in various data storehouses throughout the corporation. Thus, the agent's desktop system must be integrated with back-office and front-office systems. That's where CTI earns its pay.

CTI is the bedrock of many CRM tools. CTI connects the customer event data (such as that provided in the IVR interaction) with pre-existing customer information and campaigns that may be under way, and sends all this information to the agent with a "screen pop." This instantaneously provides the agent with helpful information specific to the current customer interaction.

With a CTI enabled desktop, the agent has at his or her fingertips automatically retrievable information that's provided via features, such as inbound screen pops, predictive dialing, call blending, and integration of data gathered from IVR and CRM systems. Resulting in not only handling the contact quickly and efficiently, but the agent also can obtain the information necessary to handle any customer transaction without the necessity of asking unnecessary or redundant questions.

When considering a call center-based CRM system with CTI features, look for one that is:

* Browser-based with HTML support. The web browser can provide access to customer information, recommendations, and knowledge management tools simultaneously.
* An Internet application architecture that has best-of-breed functionality and accesses and aggregates resources both inside and outside the corporate firewall.

* Able to use clickstream tracking, self-learning analytics, collaborative filtering, and real-time customer profiling to deliver highly targeted product recommendations, loyalty offers, and other personalized content for both web and call center visitors.
* Multichannel capable.
* Co-browsing, i.e. the ability to push to the customer and/or share with the customer any screen or content during a web-based customer service interaction.

Two areas where the CTI/CRM partnership really shines is (1) the enforcement of consistent quality during any customer contact and (2) CTI-based tools that work in conjunction with CRM systems to create easy-to-follow call guides, which support scripting, branching and legacy integration. To reduce staff training requirements, vendors have introduced products that allow non-technical call center supervisors to develop practical call guides that can lead inexperienced agents through intricate call flows.

As an example, the GartnerGroup (a well-known market research company) found during one of its case studies that the careful consideration and implementation of an integrated, multifunctional agent desktop environment resulted in a reduction in training time of 70% (from 10 to 3 days). This was attributable to easy-to-use graphical user interfaces, standard processes, guided scripting and access to contact profiles. The same company reported a 15% growth in sales by tracking and qualifying inbound calls for new opportunities, as agents could quickly demonstrate business competency and spend more time on terms and conditions. The company in the case study also found that it was able to reduce agent attrition rates by 20%.

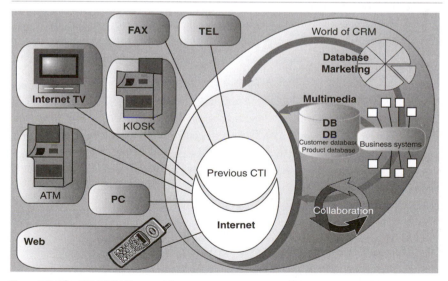

Figure 25. The CTI/CRM Partnership.

Multiple Channel Capabilities

It's common knowledge within the business community that 70% of today's business transactions are executed over the telephone and most of these transactions are conducted in the call center environment. Yet, the number of business interactions that are conducted through alternative channels is growing and as the number of customer contact options grows, call centers find that they must be able to address multiple contact channels.

CRM call center components offering multichannel capabilities that provide blending of customer contacts are extremely popular. Whether the customer opts to contact a business via telephone (including self service and voice mail), fax, email (with attachments) or the web (call-back and voice or text-based chat), the CRM system should provide full, seamless support. The CRM solution should include components that enable web chat (voice and text-based), response to a call-back request, or assisted browsing (push technology that allows an agent to send a web page to the customer's computer screen and co-browsing capabilities). Some of the more costly products might include a knowledge management engine that provides, for instance, not only:

* FAQs, but also insight into which are most queried.
* A searchable knowledge base, but also insight into the search strings.
* The ability to record all interactive web chat sessions whether with a live agent or a bot, but also analysis and parsing capabilities, which then allows insightful updates to the knowledge base.

Faxes, emails and web-based communications can overwhelm an under prepared call center. But, with CRM at the helm and good integration techniques, it's possible to store, analyze and gain easy access to the mass of customer data that flows through a typical call center's multichannel systems on a daily basis.

The following systems allow a call center to quickly and accurately respond to such communications.

Fax

Outbound fax-back (aka fax-on-demand) is usually the most common functionality associated with faxing in a call center environment. This service to the customer is typically an automated response based on a customer's IVR or voice request. Another fax technology is inbound optical character recognition (OCR) or fax-to-email and is similar to email message handling. The fax is scanned, and an email message is created.

Email Response Management System (ERMS).

Email management with auto-response technology is a fast-growing subset of CRM. Over the past few years, call centers' email volume has risen dramatically. As the need for email support in the call center environment grows, so does the cost. If not managed effectively, email support can be as costly as a telephone-based contact since agents and customers often have to bounce messages back and fourth several times in order to nail down the answer.

Fortunately, many vendors offer email routing help. Call center management already knows how to set up procedures for directing the customer to an agent who is trained to help with specific customer issues (ACDs have been doing this for years). It's the same for email routing software to distribute messages equitably among agents who have the expertise to answer them.

Experience with an IVR teaches how to offload (based on predetermined rules) some calls to a system that can automatically deal with some customer's queries. Similarly, it's possible to reduce the number of email messages sent to live agents by using auto-reply software to automatically answer email messages that only require a database lookup. As illustration: "What are your hours of operation?" "Do you take credit cards?" "Do you carry a red crew neck cable sweater in 100% cotton rather than a cotton blend?" "Do you provide home delivery in my area?" You get the idea.

It's also possible to assign specific agents to answer different types of messages (skills-base routing) or establish rules for directing messages to agents (rules-based routing). Ergo, it's possible to route email messages to agents based on skill, customer, time of day (if agents work on shifts), or keywords (whether in the subject line or the body of the message).

Some products even use artificial intelligence to identify the most likely category for each message and then quickly and accurately answer a high volume of email without human intervention. The sophisticated technology "reads" the email and prepares a response or offers a suggested response. Of course, *suggested* responses are forwarded to a specifically trained agent to determine if the response adequately answers the customer inquiry.

Minimum requirements for an ERMS system should be:

* Routing, prioritization and tracking of customer email based on content.
* Automatic acknowledgment of the receipt of a customer email by sending an immediate, standard response.
* Artificial intelligence technology to craft a response from a pre-determined list of FAQs, or to suggest answers to questions and place them immediately on an agent's screen for response.
* A coherent storage method, such as attaching a record number to each email as it comes in. Then any response(s) (including non-email responses) can have the same record number.

Many vendors offer even more functionality, such as:

* Performing on-the-fly administrative tasks (i.e. changing routing rules from a web browser). Of course, for security reasons, only authorized staff should be allowed to alter these rules.
* Classification and sorting of incoming messages. A range of automation

solutions is available for classification and sorting. Typically, the ERMS intelligently scans the message and classifies it using keywords. It may also use pre-programmed criteria such as the priority status of the sender. The latter is particularly important when there is a desire to optimize service to the "best" customers.

* Ticketing (i.e. a tracking number that allows for internal and external customer tracking). Some systems will automatically respond to a customer by email with a ticket number, giving the customer additional confidence that his or her inquiry or request is actually being addressed.
* Internet-based forms. ERMS classification, sorting and routing can be simplified through the use of Internet forms, as opposed to the customer using free-form email messaging. These forms provide for a standardized email message that incorporates pre-defined elements for identifying the subject and origin of the message.

E-Service

Web operations are rarely found within the domain of call center management. Nevertheless, to comprehend the scope of the partnership of CRM, the web and customer service, the CRM team and call center management should have a complete understanding of how a web presence offers potential savings by encouraging visitors to find answers for themselves prior to going the call center route. Many companies have attempted to move much of their customer interaction to web self-service in an attempt to reduce their overall customer service expense. At a basic level, this consists of pages showing the answers to frequently asked questions (FAQs). At a more advanced level are websites that offer vast knowledge bases, which the customer can search for answers. (These same tools should be available to the call center agent.)

As a point in fact, many customers prefer to deal with a business through an online channel, especially when seeking information on a product or service. Then the customers' web-based, self-service interactions are integrated with CRM applications, utilizing a standard SQL database.

Yet, no matter how sophisticated the on-line self-service system may be (i.e. Microsoft's website) the company must *always* provide an *easy* opt-out to a live agent (unlike Microsoft's website).

As the business community continues to move more of its selling processes to the Internet, many customers have become frustrated consumers. Consider this common situation encountered by many online shoppers: They begin the shopping process only to find themselves stymied by unanswered questions — they need real-time assistance from a live person. For instance, if I buy this vacuum cleaner, can I easily obtain replacement bags and if so, where? Or, I would like to purchase this new television set but I want to be home when delivery is made, how can I assure this will be the case? Without quick, easy answers to questions such as these, the customer will abandon the site without a purchase having been made. While self-service features can successfully solve some customer issues, others are not so easily resolved. Websites have attempted to solve this conundrum with a mixture of technologies, including the ability for customers to gain access to a live agent. For example, CRM vendors offer specialized processes and systems that make it possible to quickly connect online customers to live agents (either via voice or text chat) to obtain the assistance needed. Another option that can be added is the use of a web call-back feature, which allows the customer to ask for a live agent to call the number given, i.e. the traditional telephone is back in play. Some products even include co-browsing, where an agent takes over customer's screen. This feature is useful to help fill out a form; to send a URL for additional reference, text, multimedia or software files; to push a web page to a window displayed on the customer's screen.

The Economics of Text-base Chat

Agents who are assigned to text-based chat can communicate with several customers simultaneously. While it can be a little confusing for the agent, each text messaging session between an agent and a customer is distinct, so if needed, agents can learn to switch easily back and forth among customers.

Many companies can experience up to a 25% increase in closed sales by simply adding real-time interaction capabilities — chat, call-back — to their website. But the systems must be designed to make them familiar and customer-friendly. To be fully integrated into a call center's process and system, web-based interaction must be governed by the same standards, values and business process that govern other forms of customer contact.

As companies integrate their call centers with the web, they enable their customers to cost-effectively:

* Serve themselves.
* Schedule a call-back.
* Initiate online chat (text or voice) sessions with a call center agent.

At the same time, the company can conserve one of its more expensive and valuable resources — the call center agent - and increase web-based sales.

The use of real-time web chat for both business-to-consumer and business-to-business use is growing exponentially. Web chat capabilities should include the following:

* Historical logging, i.e. the compilation of the running transcript of an agent/customer chat session. The transcript can be stored as a customer information element or emailed to another agent or employee, should a customer issue be escalated up the line.

✳ Chat bots (similar to IVR queuing). While a customer is waiting for an initial response from an agent, the chat bot can welcome, convey information and make initial inquiries to prepare both the customer and the agent for their interaction.

When looking at e-service components, the most important tools are those that:

✳ Provide the option for a personalized self-service experience.

✳ Provide the customer with the ability to escalate to assisted service and seamlessly transfer information across all channels.

✳ Queue all forms of customer/agent interaction — voice, call-back, web chat — and provide media blending with integrated skills-based and load balancing capabilities.

✳ Consolidate knowledge bases, wherever they are located, and make the appropriate data available to customers and company personnel.

✳ Learn through customer feedback, then rapidly develop solutions to provide proactive service to the customer.

✳ Easily and quickly transfer customer history from the call center's systems to other corporate systems.

✳ Cost effectively scale as the company's service needs grow.

✳ Provide seamless integration between all technologies.

✳ Handle multiple sessions concurrently, regardless of the contact channel used to reach the center.

✳ Provide strong workflow functionality.

✳ Provide dynamic self-service and custom web page presentation through inference technology.

Document Generators

One of the most important duties of any CRM system is the nurturing and maintenance of a personal relationship with the customer. The ability for call center agents to view all appropriate information about the customer and the ability to identify products and services that accurately reflect the customer's desires is imperative. For example, some CRM systems have document generators, which enable call center agents to create personalized, "one-to-one" correspondence based upon the customer's profile coupled with information gathered during the contact. The correspondence can be created immediately and automatically after a customer contact and can be distributed via the customer's preferred channel.

Campaign Management

Integrate the call center systems into the marketing department's campaign management system for seamless communication and data sharing between the departments. This system can be used for inbound as well as outbound marketing. For instance, record selection is often used to choose random samples of customer records for a direct mail program. In the call center, record selection can respond to a customer behavior trigger. Some applications even offer "referee" functions, which can decide, for example, whether a phone, mail or email offer is most appropriate for an inbound calling customer. Look at program tracking and analysis tools, which are great for both traditional test and control group evaluations. The resulting data can be used to refine program design and selection models, and to stage the next appropriate action for customers.

Knowledge Management

Knowledge begets relationships. Customers want a company to have a corporate familiarity with them; knowledge is the key to that cognition. More important is that an informed staff provides better and more efficient sales and service, since it can better meet the customers needs. Knowledge management solutions also ensure quality in customer service interactions. The repository of knowledge that drives the system might consist of both corporate and third-party data.

These solutions provide information that can be accessed either automatically by the call center's systems or manually by a call center agent. The information typically consists of scripts and outlines that assist the agent in responding to customer inquiries. The information may also be in the form of canned email responses or HTML pages that can be pushed to the customer's browser.

A typical scenario: When a customer inquiry (no matter the contact channel) is received, an agent checks the knowledge base and customer databases for information about the customer including basic data about (say) which kitchen appliances the customer uses and product guides for the appliances. If the agent can't answer the question with the information at hand, he or she types in the details of the customer's problem and transfers it to a knowledge base where it can be used as search criteria. The agent then launches a search for a solution that answers the customer's question.

Some knowledge management systems can even allow the use of natural language statements to describe a problem through multidimensional databases and analysis. The ability to use natural language to search the knowledge base for information is a useful feature. It allows for fast access to relevant information and reduces the amount of time and effort needed to train agents to use a knowledge management system. The agent can define a problem statement as a goal, fact, symptom, change, cause or fix. For example, a symptom might be, "The X model stovetop's griddle consistently burns pancakes." A fix statement might read, "The temperature of the griddle installed in stovetop model X consistently burns pancakes. What is the best way to reduce the preset temperature of the griddle?" Such flexibility allows an agent to take into account multiple factors when searching for solutions. The system then presents the agent with a weighted list of solutions, not a tree-based, hierarchical list.

If the initial agent can't solve the problem, the call can be escalated to a more knowledgeable person — one who may be outside of the call center. The data entered during the initial call is also forwarded. Then the next problem solver can either synthesize all of the relevant information in the knowledge base into a new solution or follow up with an appropriate expert to find the answer. The new solution then becomes part of the knowledge base.

It's imperative that various points of customer interaction are integrated in such a way that any customer event is fully informed by all other customer events, regardless of the medium of communication. A knowledge management system should:

* Give agents the ability to publish and classify solutions dynamically after solving customers' problems or inquiries.
* Provide a real-time environment in which to collaborate, share and publish.
* Enable the most frequently accessed content to be identified and moved up in the data hierarchy for easy use by agents.
* Enable an agent to solve a problem once, publish it, and then allow other agents to have immediate access to the solution for other customers.
* Maintain a list or grouping of the most frequently accessed knowledge solutions (FAQs). There should be something akin to a hot key or another simple method for viewing the FAQs.
* Provide a consistent set of business rules to manage all channels.
* Be browser-based for fast, easy access to the entire knowledge base.
* Provide tools to capture knowledge both internally and externally.
* Enable full integration with call management and CRM systems.
* Be easily integrated to email and chat technologies.
* Able to connect multiple knowledge servers to the same database.

Workflow Management

Integrating workflow automation into the call center is a good move. It can help, for example, to reduce fulfillment time of new product orders while allowing call center agents to be better informed of the current status of a customer request. Progress can be monitored immediately and automatically from a desktop application — even when multiple departments are involved. As a result, customer satisfaction and associated revenues increase as requests and orders are fulfilled quickly and accurately.

Workflow automation tools can also be used to keep, for example, R&D abreast of how product problems and deficiencies are fixed and how the fix is supplied to the customer. Or keep quality assurance in the loop so management can decide when a customer issue has become a quality issue.

Call Center Management

Management needs to be able to review and monitor services and interactions with the customers on a corporate-wide basis. As pointed out throughout this chapter, the business process needed for dealing with a customer request can involve other departments. This results in a particularly irksome problem for call center management. When a customer or customer inquiry is forwarded outside the call center environment, management may have no way of assuring completion or of tracking the progress of the inquiry.

Note: Look for a product that provides an automated escalation or notification system to warn management if response times or other commitments are not being met.

Call center managers also need to know their agents' strengths and weaknesses. Which agents should be assigned to inbound (or outbound) telephone calls? Which agents can move between inbound and outbound with ease? Which agents have the skills needed to handle email and text chat communication? Are there agents that can handle text-based communication and inbound or outbound calls? This type of detail is "a given" in traditional call center technology, but ask to be sure it's also the norm in the CRM system under consideration.

Other Considerations

Most call centers already have graphical workflow mapping tools. But, if that is an option under consideration, look for products that have the ability to work in both graphical and programming/scripting modes.

Scalability is another criteria that shouldn't be overlooked when shopping for the right system. The ability for call center management to implement ("turn on") additional features as the need arises — not hiring a group of programmers every time additional features are needed — is essential

Determine the ROI. Carefully research and source CRM solutions and vendors. Conservatively budget time and resources for CRM implementation. Look at options like HSPs (see Chapter 14) and live agent outsourcing.

When deploying CRM, develop rules and processes that treat all customers well, but give exceptional service to the best. Devise and implement metrics that measure CRM performance based on customer service, such as retention, resolution and satisfaction. Then organize the call center around them. Give more weight to answering and satisfying customers, so they won't have to contact the company again on the same subject matter or have the inquiry escalated.

❂ FIELD SERVICE

Field service operations may be the final frontier in the customer/business 360-degree relationship linkage. So far, few within the business community have taken the next logical step — automating the link between field service and call center-based CRM systems. Although CRM/field service links are rare, such systems are on the rise. It's even possible to find a few vendors who offer linking capabilities whereby data generated in the field can find its way back into the CRM system to be analyzed and parsed.

Once the system is in place, field service staff can feed information about work orders, service calls and customer reactions to new products back to CRM systems for analysis. CRM systems, in turn, can provide the staff in the field with crucial information about dynamic events and customer updates gathered from all touch points and departments. For example, a customer canceled a service call at the last minute, R&D discovered a possible bug in the equipment scheduled to be serviced or an easier way to repair a product is discovered by another field service representative.

Linking field service systems to CRM systems also allow field service personnel to play a more productive role. With the proper training, they can help up-sell (e.g. new equipment, service contracts) to customers, raising the customer relationship through field service operations to a new level.

Eventually, CRM systems can be so tightly connected to field service systems that customers will be able to go online and:

*Get a diagnosis to their problem.

*Request an appointment for a service visit.

*Receive a confirmation either immediately or by email.

*Receive an email or phone call reminding them of the appointment 24 hours prior to the scheduled visit.

*Maybe even receive a notification when the field service person is en route.

Minimum Criteria

When looking at field service components within a CRM system, consider:

*Work order dispatching.

*Part order and reservation capability.

*Preventative maintenance scheduling.

*Real-time data transfer to and from wireless devices.

*The ability to track inventory handled and assigned to field force staff.

*The ability to access, record and track problem resolutions while in the field.

*The ability to address device compatibility issues.

Wireless

With field service systems and CRM working in tandem with wireless technology, field service staff can receive up-to-the-minute work orders via wireless devices, speeding up productivity, improving logistics and avoiding cancelled appointments. Such systems can result in a 25% productivity increase. Let's walk through a typical field service situation.

Without the necessity of visiting the corporate premise, the field service employee, using his or her wireless device, syncs up to the call center system. This allows the employee to download detailed information on a scheduled customer's repair order, including, for example, the exact routing to the customer's location, spare parts availability, data on any other recent repairs on that customer's equipment, and outstanding work orders including recall situations.

As technology becomes more sophisticated, the repair tasks become more complex. In general, today's field service personnel need volumes of information — more than what can be easily contained in a paper manual. Wireless devices, with an uplink to knowledge bases, and online technical manuals can reduce the time spent per call and cut down the number of revisits required.

At a tactical level, field service personnel continually return to their place of business for a trouble ticket or other paperwork. With wireless capabilities the trouble ticket or work order or whatever is pushed to their wireless device. Yet, at the realistic level, field service employees tend to resist doing paperwork and hate filling out forms. A call center-based CRM system with wireless capabilities has the ability to provide a system whereby field service employees can enter key information at the point of contact, using a handheld device that's fed short forms with simple questions. Hence, the field service group doesn't feel hassled and will fill out the forms, saving time while high-quality, valuable data is plowed back into the CRM system for corporate-wide use.

The ideal is to allow the field service staff access to forms and information they need immediately (rather than having to log on to a network). Yet, this ideal environment is challenged by the inability of many vendors to provide products that can push the necessary data onto a small screen. Most CRM vendors offer all sorts of information in a full-screen environment, but few can pull out the critical pieces and put them on a WAP or WML (wireless mark-up language), or other like device.

Device Compatibility
Many field service personnel use a variety of devices as access points, so data and forms must be consistent across all types of devices to enable a CRM-field service

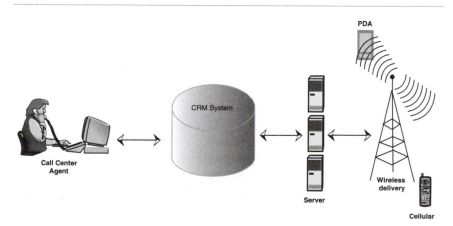

Figure 26. Wireless access to call center-based CRM systems can help field service representatives and call center agents improve customer service, speed stock and work order changes, and stabilize workflow.

linkup. When choosing a vendor select one that can handle the variety of devices the field service personnel use.

XML

Extensible markup language (XML) provides the programming and integrator communities with a way to encode and format data so it can be shared efficiently between disparate systems. The very nature of CRM systems requires that data flow freely from many sources to many destinations — XML make it possible.

Seriously consider products which use XML. XML makes it easier to standardize different types of data and XML is well on the way to bringing some standardization to the market. For instance, XML-based tools can create quick, custom solutions that can pull CRM data from anywhere for use in the field.

Note: XML is still not widely adopted and many vendors use Visual Basic or Java to create web-based pages that can be shared by CRM and field service operations.

Security

Security is a big concern and security issues abound, especially issues involved in accessing sensitive customer data from a wireless device. While most within the business community won't have top-secret information going back and forth, the information feed still must fit in with firewalls and the corporate infrastructure.

One solution to the security issue would be to find a vendor that can provide field service data in an HTML-only version. Then, the server can act as an automatic barrier.

Alternatives

Wireless still faces hurdles — sorting through wireless protocols and standards, as well as ensuring uninterrupted wireless coverage. For companies whose systems for field service operations lack good wireless coverage and/or fast data transmission speeds, an alternative is to use a Windows CE-based system and supporting device. This allows the employee to store and forward data via scheduled updates.

Have a backup system. Even with an all wireless system, Ethernet, a telephone dial-up option or other backup system should be in place. If Ethernet or dial-up is available, it should be considered the option of choice when performing data synchronization — both are less costly than wireless.

Advantages

With a top notch CRM-field service system in place, a company can realize more than a 20% gain in productivity, which can usually be ascribed to an increased understanding of customer needs and better case-by-case personnel deployment.

Integration

To make CRM effective requires more than buying or contracting for software and fitting out a call center. At the risk of being redundant — corporate-wide integration is the key. Departmental silos must be breached to unify customer service elements. Install CRM in such a way that service is layered across different business functions. All call centers provide data that can aid a business in the development of new products and campaigns and refinement of existing products and marketing initiatives. As illustration:

* The existence of most marketing departments' is dependent upon the data provided by call centers.
* Many outbound call centers are divisions of the sales department.
* Accounting is beholden to call centers to field and even solve billing issues and provide credit collection duties.
* Order processing and fulfillment would be in a world of hurt without call center agents to resolve their numerous customer issues.
* Field service departments are increasingly dependent upon the call center to manage service calls.
* Research and development is indebted to call centers for the input received about design issues and product bugs.

✪ CALL CENTER STAFFING

When implementing CRM in a call center, businesses wrestle with a number of challenges, staffing being one of the most vexing. It has become increasingly difficult and costly to recruit and retain qualified staff, and the associated training costs can be overwhelming, since inexperienced agents may not be able to provide the level of service that today's customers demand.

A CRM system offers a twofold benefit: It can be used to help the call center's staff to work more productively and efficiently, and it can also help to create a better working environment (which in turn leads to better customer service). Integrating CRM within the a call center ameliorates the difficulty and cost of recruiting and retaining qualified call center agents. By automating as many processes as possible — training, call routing, and even basic call answering — it's possible to save up to 70% on staffing costs.

In addition, to help enforce consistent quality and to reduce training, some call centers benefit from tools that create easy-to-follow call guides. Vendors offer products that enable the average call center supervisor to develop sophisticated call guides that can lead inexperienced agents through complex call flows.

✪ COSTS AND BENEFITS

A *new* CRM-enabled call center can cost into the low millions of dollars to set up and maintain (*upgrading* established centers should *cost much less*). This amount includes quality, high-paid agents who can cross- and up-sell and resolve problems, not just recite information. Assuming per-agent wages and benefit costs of $30,000 annually, a firm must pay $3 million yearly to staff a call center with 100 full-time or full-time equivalent (mix of full- and part-time) agents.

Cost savings derived from self-service technologies reduce the agent/customer cost ratio. However, use of self-service techniques is not a justification for offering less than top quality service. According to observers, too many companies abuse self-service products by shunting low-priority customers through complicated menus and prompts, with no opportunity to contact a live agent. It can't be stressed enough — *always offer the option to contact a live agent.*

Companies that invest in a CRM strategy and its enabling technology within their call centers stand to reap huge rewards. Call center-based CRM can reduce the cost of sales and customer service while increasing revenues through improved cross-selling and up-selling capabilities. By gathering high-quality customer information, companies can improve their CRM metrics.

My experience and data indicate that the typical 21st Century call center's cost per customer contact breakdowns something like this:

* Close to $35 per inquiry if received via toll free telephone.
* $5 to $7 if some form of web chat is used.
* $5 to $10 per inquiry if both inquiry and response are email based.
* Web and IVR self-service average just 25 cents and 55 cents, respectively.

Forrester Research, a leading independent research firm, agrees. A recent Forrester Research study calculated the average costs for a call center session at $33, whereas on-line chat runs at around $7.00 per query and access to a knowledge base via a web-based query costs at close to $1.00.

The numbers tell the tale. Companies willing to invest in a CRM strategy and enabling technology within the call center environment reap huge rewards:

* Reduction in the cost of sales and customer service.
* Higher revenues and greater profitability.
* Better customer data.
* Improved employee morale.
* Improvement in the overall quality of customer interactions.
* Increased customer loyalty and satisfaction.

Want more? With the right CRM tools onboard, a call center can see an agent's productivity increase about 15% (quicker call resolution and higher rate of resolution during first contact) while at the same time achieving at least a 30% reduction in training costs. Conversely, as indicated by numerous industry surveys, more than half of the customers who quit a business cite inattention as the cause (less than 5% of that group can be enticed to return).

More importantly, companies can improve the overall quality of customer interaction while streamlining customer requests and orders. As a result, businesses can achieve increased customer loyalty and satisfaction and ultimately greater profitability when they adopt a CRM strategy.

In addition, corporate executives discover that by equipping the call center with the proper tools, employee attitude and morale improves due to reduction in their level of frustration. This translates into better customer service and an improved profit margin.

⊛ IN SUMMARY

The high cost of direct sales, along with increased customer demand for convenience of interaction, is forcing changes in customer relationship strategies; and the call center is emerging as a strategic part of the relationship management effort. Today's, call centers are increasingly responsible for business interactions conducted through alternative communications channels such as email, the Internet, fax, and voice mail.

Many top-level corporate executives fail to take note that the biggest advantage of having state-of-the-art CRM capabilities within a call center environment goes to the *customer* — a business's most valuable asset. They ignore that when a customer finds himself or herself being forced to ask the same question of six different employees and getting three or four different answers, or gets caught up in the "bounce syndrome," that customer loses confidence in the product or service and, finally, the company itself. With a CRM strategy in residence, that won't occur — everyone throughout the company has access to the same data. Customers greatly appreciate a business that can give them a precise response to their inquiries the first time — no matter how thorny the issue might be.

While the technology described in this section is not intended to provide the details of an all-encompassing template to build a state-of-the-art call center, it should provide the foundation necessary to start a dialogue within the company. That dialogue may be the first step in gaining the executive board's attention to the need to differentiate the business competitively through optimal customer service. Whether the business operates out of brick-and-mortar locations, through catalogs, or over the Internet, it is spending money to acquire its customers. That money is wasted if the call center's systems fall short of delivering exceptional customer service.

Note: For complete details on establishing and upgrading call centers, pick up Brendan Read's book "Designing the Best Call Center for Your Business" and Andrew Waite's latest book, " A Practical Guide to Call Center Technology: Winning and Keeping Customers by Getting Calls to the Right Person."

Chapter 13
The Importance of Data

The objective of CRM — understanding and managing customer relationships — depends on the proper integration of a wide variety of data sources.

While the many flavors of CRM technology and IT infrastructures make it difficult to describe a "typical" CRM data architecture; the basic data architecture for any CRM initiative includes a data storehouse (whether it's a database, date mart or data warehouse) as the central hub. This repository of data absorbs information from sundry sources and then provides interaction internally and externally.

Typical CRM-related Data Repositories

Databases: Typically found throughout the corporate environment. A database is a set of related files that is created and managed by a database management system (DBMS) or if a relational database by a relational database management system (RDBMS). The databases that a CRM system can most effectively use are the relational databases. In non-relational systems (hierarchical, network), records in one file contain embedded pointers to the locations of records in another, such as customers to orders and vendors to purchases. These are fixed links set up ahead of time to speed up daily processing. In a relational database, relationships between files are created by comparing data, such as account numbers and names. A relational system has the flexibility to take any two or more files and generate a new file from the records that meet the matching criteria. Routine queries often involve more than one data file. For example, a customer file and an order file can be linked in order to ask a question that relates to information in both files, such as the names of the customers who purchased a particular product.

Data Marts: Typically found operating in conjunction with sales and marketing automation software. A data mart is a data warehouse that covers a single subject — essentially a separate database fed by operational systems and built solely to warehouse all the data that is collected to support a specific CRM initiative. Data marts contain data tailored to support the specific analytical requirements of a given department or business function that utilizes a common view of the data. Thus providing more flexibility, control and responsibility. Data marts may be set up to perform simple querying and reporting, to enable trending and multidimensional analysis, to support sophisticated data mining such as pattern recognition, or to allow unstructured exploration of the data.

Data Warehouse: Typically found within an enterprise-sized business that uses a more developed form of customer-centric marketing or managed service and support, requiring a greater integration of data, from both front- and back-office sources. The data warehouse serves as a single repository of data and usually replaces multiple data marts. Once the enterprise data warehouse is loaded with cleansed data, it provides all the data feed and receives all the data streams for the entire enterprise. Then the appropriate query, analysis tools and data mining software are integrated to allow the company to parse the data in such a way that it can be used to better understand its customers.

Integrated CRM Solution: A system that incorporates both data warehousing and e-commerce technology. It normally involves the integration of e-commerce and e-business systems with both front- and back-office systems, which then feeds and pulls data into and from a data warehouse.

Operation Data Store (ODS): A collection of data used to support the tactical decision-making process. It's subject-oriented and fully integrated (only one view of the customer), holds current data and is volatile. For marketing, the ODS generally has all of the customer's current information centrally available, e.g. customer's address, telephone number, email address, age, household income, product purchases, product utilization, etc.

Although CRM is a business strategy, it's enabled by technology. Widespread storehouses of customer data residing in ERP systems, sales force automation, call centers and integrated point-of-sale systems have made customer data available in large volumes. Database management platforms enable the business community to access this data to gain new insights into their customers through a variety of analysis methods. Perhaps most significant, the Internet provides a completely new way for businesses to interact with their customers, producing a landslide of fact-filled customer-related datastreams.

Many CRM systems are developed independently, meaning there must be expensive integration efforts to achieve the vision of true customer relationship management on a corporate-wide basis. Integration of these disparate systems is the only way to have a successful deployment of CRM corporate-wide since it's the only way to provide consistency within the customer experience. Yet, a well-defined CRM strategy and a supporting information and technology ecosystem *can* consist of point solutions *if* they are implemented with an eye toward the goal of a corporate-wide CRM strategy. With the CRM strategy plan at the helm, the pieces can eventually come together to create a unified whole with customer information flowing freely across functions and through channels.

Without a defined CRM strategy, independently developed point solutions based on function-specific short-term needs spring up like wildflowers. The marketing department implements a variety of products (sometimes combined with integrated suites) to help with the planning, execution, and monitoring of marketing campaigns. The sales department brings onboard a lead management system. Supply chain management systems and product delivery systems are deployed to support customer based customization and to provide the current location of goods in transit. Call centers deploy sophisticated multichannel support systems for ongoing customer service. While it's true that these separate CRM projects, working in their isolated departments can provide:

* A means to support function-specific and channel-specific CRM strategies.
* A business culture that can shift from product-focus to customer-focus.
* Sales and marketing that can focus on retention and increased share of customer instead of just acquisition and market share.
* Customer service within a multichannel customer environment.

They can't provide free flowing customer information. The data is trapped in departmental silos. Each department performing their tasks without truly understanding that the intelligence they are generating in their stand-alone data storehouses could be useful to others in the company. This means that no one has a full picture of who the customers are, which ones are valuable, which ones are unprofitable, what products they buy, or how many times they were contacted.

To achieve the vision of CRM these silos must be torn down. There must be integration of the various point solutions. Only through integration of marketing, sales, customer service, ERP, accounting, order processing, etc. is the vision of CRM realized.

Note: In a Global Benchmarking Survey taken in 2001 only one company had full data integration of the databases which contain customer information; and, unbelievably, 17% reported no integration at all. The remainder integrates their data by function or by related database group.

Every customer-facing employee needs access to the latest information on the customer's profile, behavior, and expressed needs. This means marketing should provide promotions and offers to individual customers based on their interaction on (say) the corporate website yesterday, a conversation with a call center agent earlier today, a chance discussion with a field service representative two hours ago. Products are customized to meet specific customer needs and the customer and field service departments are informed, resulting in a more than satisfactory customer experience. With a corporate-wide view of each customer, the value of each relationship is measurable and each relationship is managed based on this value. Every customer contact becomes an opportunity to modify customer behavior in a beneficial way based on the totality of information.

⊙ THE CRM DATA ARCHITECTURE

The starting point for implementing a CRM data architecture is the conceptual data architecture. It's composed of three distinctly different sets of systems as discussed throughout this book — operational systems (supporting business operations), analytical systems (supporting business intelligence) and collaborative systems (supporting business management). All organized in an all-encompassing architecture to support the CRM initiative.

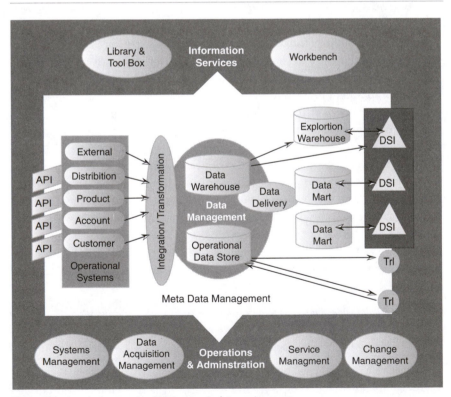

Figure 27. A typical enterprise CRM system's data environment.

Business operations are the primary producers of data and they run the business's day-to-day operations. They usually require manual interfaces with other business operations systems and are usually legacy technology — meaning limited documentation and an inability to change. Examples include order entry systems, accounting and inventory management systems.

Business intelligence provides the ability to analyze data and information used in strategic decision support. These systems are major consumers of data and consist of the data warehouse, data marts, the data storage system (DSS) interface and the processes to "get data in" and "get information out." The data warehouse is a source of integrated, re-engineered, detailed snapshots of data. Its data is prepped and cleansed within certain parameters and then documented in terms of its sources, transformation rules, calculations, and other characteristics.

The data mart is a subset of the data warehouse. Its data is formatted for a particular function or department (typically the marketing department) and is usually aggregated or summarized. The data mart normally has a known set of requirements and reports. A data mart's strategic function could include customer demographic profiling, recency/frequency/monetary (RFM) analysis, and marketing campaign planning.

Business management supplies the ability for corporations to act upon the intelligence obtained from their strategic decision support systems in a tactical fashion. It's not only a major consumer of data, but also may be a producer of data. The Operational Data Store (ODS) is the major component used in business management. The ODS is an integrated, cleansed, dynamic and current set of data for tactical decision making. It is accessible from anywhere in the organization and generally does not support any particular operational application. The ODS has a transaction-oriented interface and data structure. An example is customer and product data necessary to support websites and call centers.

When the data is consolidated through a CRM architecture approach, it gives the company and its employees a fully functioning technology setting from which to perform all the analyses, trending, and statistical studies they desire while allowing the enterprise to benefit, and then act upon, all this garnered intelligence.

Note: *Because of the amount of data flowing through CRM systems, a full-time database administrator is usually responsible for maintaining the data storehouse(s).*

Business intelligence requires a lot of detailed, historical data, which generally is static, i.e. it requires minimal transaction processing, but does require substantial analytical processing speed. Often the results from one capability are used as the input into another capability, e.g. the high lifetime value customers identified in a customer value analysis may be fed directly into a purchasing behavior analysis to determine what products this segment is buying and using. The results of that analysis then can be fed into a sales channel analysis to determine the best sales channel for reaching this valuable customer group.

The company then needs means to act upon the intelligence garnered from these analyses. Once a trend has been spotted or a behavior analyzed staff must be able to respond to that information. There needs to be an architecture that supports not only the myriad capabilities necessary for marketing, sales and customer service, but also allows for the productive, efficient, and flexible usage of the corporate data — an architecture that can support sophisticated CRM systems.

❂ THE DATA STOREHOUSE

The data warehouse plays an important roll in the delivery of strategic data as does the operational data store (ODS). While the data warehouse does share some characteristics with the ODS, there are numerous differences, for example, an ODS is:

* Subject-oriented around major subjects of data, such as customer, product, transaction, etc.
* Integrated, meaning it has one and only one version of the customer and a common view of the corporate-wide data.
* Current, at least as current as today's technology will allow.
* Volatile, which means it's capable of receiving changes and updates to its data.

Compare that to a data warehouse, which although it's subject-oriented and integrated, is also composed of:

* A series of historical data snapshots, each dated and accurate as of that point in time.

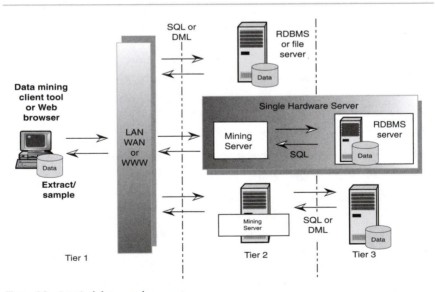

Figure 28. A typical data warehouse system.

* Non-volatile data, e.g. it cannot accept transactional updates or changes.
* A collection of static data used to support the strategic decision-making process and is the central point of data integration for business intelligence.

The most basic component of a data warehouse is a relational database where the data is stored. These databases are designed to efficiently insert new data and locate existing data using a standardized query language. Underneath the database is a maze of connectors and transformers which connect the data warehouse to other systems containing data — call center systems, SFA systems, marketing databases, accounting, etc. Of course, there is usually a need to "treat" the data before it's moved to or from the data warehouse.

A New Data Architecture

Now, new tools and technologies are available that can provide on-line transaction processing (OLTP) capabilities, ODS capabilities and data warehouse capabilities in the same database. While most OLTP products are proprietary in nature and limited to niche-sized markets, look for these capabilities to gradually become widespread and mainstream.

The systems running today are combining near real-time data feeds, heavy user loads, detail-level transaction records, extensive transaction history, analysis and trending functionality, gigabytes of data, OLTP field-level update/edit functionality and OLTP response time. These systems have eliminated four separate RDBMS

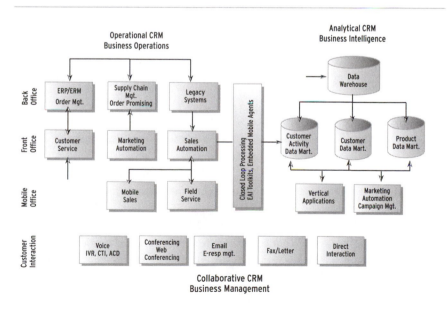

Figure 29. Typical CRM system and its data architecture.

instances (LTP, ODS, data warehouse and data mart); the extraction, transformation and loading (ETL); and replication of data across the four instances.

But what do you call this data architecture? It's not a data warehouse because of the OLTP and ODS capabilities. Data management guru, Douglas Hackney, has suggested a new title, "online information processing" or "OLIP," to define this new space.

Meta Data

Meta data is data that describes other data. In a CRM data architecture, meta data refers to any file or database that holds information about another database's structure, attributes, processing or changes — a common occurrence in CRM systems. Meta data provides the necessary details to promote data legibility, use and administration. It is the glue that holds the CRM data architecture together. It supplies definitions for data, calculations used, information about the data's source (where it came from), what was done to it (transformations, cleansing routines, integration algorithms, etc.), who is using it, when they used it, the quality metrics, etc.

Quality is critical when current data is used for tactical decision-making. A meta data environment should:

* Establish data quality baselines.
* Report on data quality metrics.
* Report on known sources of bad data.
* Establish data quality procedures (i.e. when to reject a record, how to fix a problem, when to escalate a problem, when to stop processing, etc.)

Data quality that's quantifiable can be improved. When time and resources are allocated to data quality issues, decision-makers are more confident in the data and the analyses it provides. Therefore, they will be more likely to use the system.

Data Cleansing

Gathering and storing customer data so it can be efficiently utilized is the key to CRM technology. One of the bugbears of any CRM implementation is data — especially data cleansing — but when data must be gathered from divers sources into a single data storehouse, data cleansing is a necessity. Most of the time this task must be completed before the integration process can begin. If processes are performed using uncleansed data there will be intermittent failures. If uncleansed data is fed into a data storehouse it has the potential to corrupt the data storehouse as changed data is reproduced from the source database over time. Reports that result from both cleansed and uncleansed data will not be equivalent — causing confusion and lack of trust in the results.

If it's impossible to clean the data prior to integration, a team should be put in charge of assessing the data's completeness and accuracy to determine its reliability and what improvements need to be made to assure quality data is fed into the CRM systems. This assessment must address accuracy (i.e. the valid value is a "correct" value) and not just validity (i.e. the degree to which data conforms to the business rules — is it a valid value, within the right range, related to a valid referenced object, etc.?). Data often has a high incidence of valid values — especially default values — which are valid but not correct.

Most companies cannot go back in time to improve the processes that have left them a *legacy* of defective data. So, how much should a company invest in scrapping and reworking data to correct its defective data? To get at the right answer determine who will use the data and for what purposes? What are the costs and risks of process failure or decision failure due to nonquality data — customer alienation and loss of customer lifetime value, lost and missed sales opportunities, or what? Once the answers are in hand, it's possible to determine what data needs correction and how much effort and money should be expended.

Quality Assessment

Data quality has less to do with zero-defect data than it does with consistently meeting internal and customer expectations. Most people can live with some degree of omission and error if they know both that it exists and the level of its reliability. This requires quality assessment. Data without this assessment should be labeled "unaudited data" to indicate there has been no quality assessment.

* *Priority data* should be close to zero defect (e.g. an error could have high cost of failure). For example, where the misspelling of a customer's name or incorrect address could lose a customer.
* *Important data* comes second. E.g. demographics are vital for accurate marketing campaigns but errors in the data won't cause customer attrition.
* *Non-priority data* is optional or non-critical data where the cost of omission and error is marginal. A good example is duplicate addresses.

The company *must* address cleansing of all priority data. Then address important data as resources allow. To get an idea of the extent of any accuracy problems, begin with a small sample of 50 to 100 occurrences of the "priority" and "important" data groups and measure it for accuracy. If needed, use customer surveys to assure and correct vital data.

Data Cleansing Tools

With some internal data it's possible to use data cleansing tools. These tools will not only standardize and validate the data, but also can be used to derive new data providing additional capabilities to analyze the company's customers. Let's look at basic customer data and use it as an example of what data cleansing tools can do.

* Based on a customer's address field(s), which probably includes various addresses, these tools derive (or lookup from their internal tables) geographic data such as the ZIP+4 code.
* From the customer name these tools can derive the sex of the individual

(along with a probability of certainty). For example, the tool will determine that Cliff Perciavalle is a male with a 98% certainty.

✳ The leading data cleansing tools can perform a function called householding, i.e. they can determine which customers live in the same household. Such data can save on mailing costs.

✳ Some address cleansing tools can apply address corrections by using, for example, postal service change-of-address data.

Human Intervention

There should be staff on hand to measure, confirm and correct most data to accurate values. For instance, people move without providing forwarding addresses. Someone must confirm that an address for a specific person is correct.

Dealing with Redundancy

Matching of data redundantly stored in disparately defined databases is a painful data cleansing problem. If there are redundant databases that contain records about a single customer and those records have embedded meaning or non-synchronously defined keys, the first step is to consolidate the duplicate records into one database. Keep a cross-reference table to relate the surviving "occurrence of record" to the records that previously existed but no longer do. This is used to redirect any business transaction using "old" identifiers to the occurrence of record. Also maintain an audit file with before and after images of the data to assure the original records can be reconstructed in the event of an error.

The next step is to de-duplicate and consolidate the records. Once there is a single occurrence of record, match and consolidate them across the redundant files selecting the most reliable values. Correct and synchronize data values at each source for consistency. For example, use full names as opposed to only initials and spell out street names instead of using abbreviations. This is required to maintain "equivalence" of data. Maintain a cross-reference table of related occurrences, again so consistency across the files is maintained.

Data quality has become a major point of concern in the development of sophisticated CRM systems. We've all heard the expression "garbage in — garbage out." Well, it's as relevant today as it was 25 years ago. The difference between a good query and a bad one is the amount of thought and analysis put into the query beforehand. The difference between a good response and a bad one is the quality of the data.

Lack of data quality results from operational systems doing their own processing, each with their own data, and with their own way of doing things. The quality of the data can be greatly improved simply by extracting it from the operational environment, running it through a cleansing process, integrating it with data from other systems, and transforming it into a corporate standard. All of these activities can occur in a meta data environment or a data warehouse environment or a mixture of both. The better the data cleansing process, the higher the quality of the data environment.

Data Access and Analysis Tools

Every data storehouse provides a variety of tools that can turn its data into useful information. The data access component of most CRM data architectures is comprised of several related technologies that address data access and analysis. For instance there are query, reporting and analysis tools, which offer functions ranging from basic query and report capabilities to multidimensional DBMS OLAP or relational OLAP tools for creating online analytical processing.

These tools are useful when there are many possible relationships. For instance, a durable goods company may track 100 variables about each customer. To say the least, there are scores of possible relationships among the 100 variables. Data mining tools can help point out the significant relationships. But, many times this requires multiple technologies. For example, for the decision maker there must be a *decision support interface (DSI)* that's intuitive (a person with insight and perception should be able to run the tool with minimal training) and easy to use with built-in queries or algorithms. This tool should also have the ability to interface with other applications (word processing and spreadsheets) and be able to handle massive amounts of data with ease, then display all information in a comprehensible mode.

There are also interfaces for the ODS, which are referred to as the *Transaction Interface or the TrL*. This tool can source data from operational data stores. It provides an easy-to-use intuitive interface to request and employ business management capabilities.

These varied tools fit specific user skills and needs. The most powerful need the skills of an experienced corporate analyst, while the more basic are designed for less technical information users. However, even with the more basic tools, these users (mainly managers and executives) will need training in analytics, such as how to form tactical and strategic queries. Then training in the usage of the tool itself.

Most of the tool sets that come with a data storehouse provide (with adequate training) easy exploration of the contents of the data storehouse and can test out various hypotheses. But, they depend on the user to guide the investigation. They require a predefined starting point — a hypothesis, a query, a procedure, a program — that dictates the nature of the data analysis to be carried out by the tool. These tools, alone (without insightful input from the user), can't address the very real business need to uncover previously unknown business facts.

Data Mining Tools

The business community is drowning in data while starving for real information. This conundrum has led to a widespread interest in data mining. To glean value from the information contained in the CRM data environment requires data mining/modeling tools and tools that provide multiple forms of analysis and query. Tools specifically designed to identify significant relationships that exist among variables. Yet, most companies just focus on data entry; they don't consider how they will extract and obtain value from that data. The question that begs to be asked is "If you can't get the data out of the system and utilize it, why bother collecting and entering the data in the first place?"

The business community is beginning to see the light. Companies are beginning to use a series of data mining tools to help them predict future trends based on an analysis of historical behavior.

Data Mining

Data mining is the process of selecting, exploring and modeling large amounts of data to uncover previously unknown patterns for business advantage. This enables a company to produce new customer intelligence. Data mining can use data collected from a variety of sources, including corporate transactions, customer histories and demographics, even external sources such as credit bureaus. Then data mining tools using statistical and machine learning techniques can take the results and predict customer behavior and produce patterns in the data that can support decision making and predict trends. In this way new selling and/or customer enhancement opportunities (for example) can be created.

Data mining technology is the most recent addition to the data access component field. Data mining tools bring the practice of information processing closer to providing the real answers to problems businesses seek to address from using their data. Many industry experts think data mining can supplement human management insights, allowing businesses to make more proactive, knowledge-driven decisions in their quest to remain competitive.

Note: *Data mining results are highly dependent upon quality data. Their techniques are highly sensitive to missing or inconsistent data.*

Data mining tools typically address two key business requirements: First, is the use of patterns to predict future trends and behaviors (i.e. what's the likelihood the customer will buy this product?). This is referred to as the *prediction* requirement. The second is the ability to discover patterns, associations and clusters of information (i.e. spending patterns). This is referred to as the *description* requirement.

There are six major categories of data mining methods that these two business requirements use. The prediction requirement primarily uses:

* Classification. The development of profiles of groups of items in terms of their attributes.

✳ Regression. The establishment of a relationship between a series of items for the purposes of forecasting.

✳ Time series. Similar to regression but this method uses additional properties of timed information.

The description requirement primarily uses:

✳ Clustering. The segmentation into subsets or clusters items that exhibit consistent behavior or characteristics.

✳ Association analysis. The recognition that the presence of one set of items implies the presence of another set of items.

✳ Sequence discovery. The recognition that one set of items is followed by another set of items.

There are numerous data mining techniques, with more being discovered every day. But, the most commonly used data mining techniques are:

✳ *Decision Trees*. A tree-shaped structure that visually describes a condition that caused a decision to be made. Rules then can be generated for the automatic classification of a data set. A couple of common decision tree methods are classification and regression trees (CART) and chi square automatic interaction detection (CHAID). Decision trees are often combined with neural networks (non-linear predictive models that learn how to detect patterns). They can also be combined with rule induction (the process of extracting useful "if...then..." rules from data based on statistical significance).

✳ *Genetic Algorithms*. Optimized techniques that can be used to improve data mining algorithms enabling them to derive the best model for a given data set. The resulting model is then applied to data to uncover hidden patterns or to make predictions. Genetic algorithms are best suited to segmentation/clustering applications.

✳ *Predictive Modeling*. A variety of techniques used to identify patterns that can then be used to predict the future, such as linear regression, logistic regression analysis, generalized additive models (GAM) and multivariate adaptive regression splines (MARS). These techniques are most often used to predict the odds of a particular outcome, based upon the observed data.

✳ *Fuzzy Logic*. This technique handles imprecise concepts and is more flexible than other techniques. Direct mailers commonly use this technique.

✳ *K-Nearest Neighbor (k-NN)*. This technique places an object of interest into a class or group by examining it with others whose attributes is closest to it. This technique is used for discovering associations and sequences when the data attributes are numeric.

✳ *Neural Networks*. This technique is appropriate for clustering, sequencing and prediction problems. However, neural networks don't explain why a particular conclusion was reached.

Selection Criteria

When considering a data mining tool keep in mind that to make the best use of data mining techniques the tools should be integrated with a data warehouse. In the quest for a data mining tool the data team and/or CRM team must determine:

* *Its architecture*. The data team should understand its operational and connectivity requirements and how its data (and meta data) is stored. Find out if it uses sampling techniques to process a representative data subset or a direct access technique whereby it processes the data by accessing the data storehouse directly. Note that direct access tools that read the data using the native SQL of the database server are likely to be more scalable as data volume grows. But also be aware that a sampling approach will usually offer more opportunity for data cleansing and other data preparation activities.

* *Integration abilities*. In other words, how well does the tool integrate with other components within the CRM data architecture. Also determine where and how the meta data is stored. Ask about rule conversion, it's an important requirement for the integration of data mining tools with other data analysis applications such as OLAP. If it can convert the rules it has discovered into (say) SQL code, then other decision support tools can reuse that code. Also determine if it can interface with an OLAP tool, thus allowing a two-way exchange of information.

* *Performance*. Can the tool exploit multiprocessors by running, for example, multiple instances of itself, or by handling the processing for a mining technique like a neural network in parallel?

* *Function*. How many different data mining approaches and techniques does the tool support? Does it provide a set of predetermined rule models and applications?

* *Presentation capabilities*. A rich set of data visualization capabilities is important. So, determine if it supports data visualization. Ask if it provides different ways to view the mined data. Is it intuitive and does it integrate seamlessly into the user's current technical environment? Can it interface with workflow and other necessary applications?

* *Data sources*. The data teams should evaluate the level of support for various possible data sources. For instance it may be important to use external data sources in the mining process.

* *Data preparation*. Does the tool provide support for data preparation and cleansing?

* *The environment*. What client and server platforms does it support? Are there any limitations on the size or amount of data that can be mined? Most of today's data mining tools tend to be available for only a limited number of data environments and platforms.

* *Vendor resources*. Determine if the vendor has the resources to maintain future evolution of the tools?

✳ *Training.* Determine what type of training is available and if there is additional charge for the training.

The company will probably need to deploy several different mining tools to satisfy the full range of the CRM initiative's data mining needs. For example, there will probably be a strong interest in web-enabled data mining.

The leading data mining tools are more than modeling engines with powerful algorithms, today's data mining products address the broader business and technical issues, such as integration into CRM's data architecture where the mined data is fed directly into, for example, a campaign management system. The current class of data mining tools can model virtually any customer activity by finding hidden patterns that experts may miss because it lies outside their expectations, then present the information in such a way that's relevant to current business problems.

❂ THE WEB

Customizing the individual customer's web experience is becoming common place. But for such customization to occur data must flow rapidly through the data architecture to the points of analysis.

The knowledge and skill to cleanse web-based data is a developed rigor. The data architecture must find threads through the data that are correct and associate the web-based data to all of its dimensional properties in order to make it usable. Unactionable web-generated data must be filtered and then a subject-oriented data model parses this mass of data with its complex structures and relationships to ensure that the data has business meaning. Although a chore, it's worth the effort. Marketing professionals actually salivate over the gigabytes of customer data collected by corporate websites. This data is chock-full of interesting customer tidbits that can be used for trend analysis, customer analytics and marketing campaigns.

❂ MANAGING DATA IN THE MID-SIZED BUSINESS

In the enterprise environment, the division between the major systems of the CRM data environment is fairly distinct. However, in the less than enterprise-sized business, the distinction is less clear. A small business may not need separate business management and business intelligence systems. Many CRM capabilities may be implemented using purely operational systems and the information technology components of business operations together with PC-based spreadsheets, databases and contact management systems.

But, one constant remains even for this business group, the CRM systems still must be unified, integrated with a data architecture that spans the CRM ecosystem. Even for this market, CRM capabilities are limited if there is only a limited view of the customer. Corporate-wide customer-centric focus isn't possible if, for example, inventory status isn't available for all that need the information; customer dissatisfaction isn't communicated to all departments; or customer touch points aren't pulled together so there can be a comprehensive understanding of the customer. The customer suffers — there is an inconsistent customer experience.

✪ KNOWLEDGE MANAGEMENT

The knowledge or business intelligence accumulated throughout a business's IT infrastructure on subjects ranging from customer behavior and industry trends to detailed product information and quality issues and flowed into a data storehouse(s) is manna for management. They use this data to grow revenue, build long-term customer relationships, and shape products based on emerging trends.

Knowledge management can help management to spot winning and losing products. It allows constant monitoring and review of customers' response to marketing and sales efforts, to products and services. Knowledge about customer preferences and expectations can be used to make proactive changes, enabling a company to continuously improve existing strategies and to formulate new strategies. Knowledge management within a CRM architecture also gives management the ability to analyze a business's strengths and weaknesses (e.g. find out the impetus for a sale and conversely, the reason why a sale was lost) — allowing maximization of a business's strengths and the means to address its weaknesses.

Knowledge management entails the capture, processing, storage, delivery and maintenance of huge amounts of data about products, people, processes and policies. Much of the data is generated in the workflow, as employees discover the facts that need documenting; other data comes from diverse sources. All data comes in diverse formats. Sorting out who needs what knowledge, when, in what context, in what form, and for what purpose, is a strategic challenge in itself, i.e., "Where does this data come from?" "How does it get into the system?" "What can be done with these terabytes of data?

The king of knowledge management tools is data warehousing (but ODS is no slacker), which can gather information on the ever-changing needs of a company's customers and marketplace. Coupled with the use of operational, analytical, transactional, interactive, modeling, historical data, and predictive knowledge processes, data warehouse technology can effectively convert the gathered data into useful knowledge. The intent is to enable a management team to focus on the results of an analysis of business improvement alternatives rather than on the mechanics of the analysis and the associated difficulty involved in identifying, compiling and sorting through all the variables.

For instance, technology in the form of advanced techniques of data transformation and graphical presentation can accelerate the usefulness and speed of management decision-making. Other systems can help managers to analyze available data, so as to highlight trends, and prepare suggested courses of action. All of which can help in the decision-making process.

Customer Intelligence

True customer intelligence is dependent upon data warehousing technology, which ties front- and back-office systems together to track customers from the first contact through to the current contact information.

For example, customer intelligence solutions offer the only way a company can implement campaign management. Campaign management tools leverage data in the data warehouse to create targeted, intelligent marketing campaigns.

◎ IN SUMMARY

Although most businesses recognize data as an asset, they don't have the technical wherewithal to treat it as an asset. They can't exploit their data assets without implementing new processes and perspectives on data.

To fully exploit data, the business community must acknowledge the broad spectrum of data that exists and it must incorporate new data management principles (such as CRM) into all business efforts. Businesses also must overcome the corporate attitude of rebellion against the combining of data processes with business processes. Finally, meta data must evolve within the corporate data environment.

The bottom line: Data as it is know today is evolving outside of its traditional role. Thus, the tools and philosophies of the past are no longer adequate.

Chapter 14
Hosted Solutions

The urgency within the business community to adopt or expand CRM functionality has caused a surge of interest in the hosted solution model. Yet, entrusting the technology end of a CRM strategy to a third party can be a little scary. Through adoption of Internet standards, strong partnerships and the sharing of best practices, the hosted solution industry is working hard to address concerns about security, availability and control.

Due to this shift in corporate thinking, the hosted solution model has become more appealing in the corporate boardroom. Many corporate executives are starting to wonder, "What if someone outside the corporate walls could provide, maintain, and update this technology for us?" Even IT executives, who traditionally have felt that they needed to own all their applications and keep them all in-house, are beginning to rethink their stance. Large- to enterprise-sized corporations that own a multitude of best-of-breed applications with all the related service, support and management headaches are also perusing the hosted solution marketplace.

There are many flavors and choices available in the hosted solution arena. As with an in-house solution, the CRM team needs to understand their company's CRM goals and objectives, before going with a hosted solution.

✪ WHAT IS A HOSTED SOLUTION?

For the purpose of this book, a hosted solution is, basically, a new model for selling software made possible by the marriage of the Internet and conventional CRM products. A hosted solution provider (HSP) hosts software applications. These applications are stored and managed off-site at a data center either owned or contracted for by the HSP and provided to the client companies over the Internet or Intranet via a per user, per month pricing model. Thus, the HSP provides its client companies with all the necessary hardware and software, a secure network infrastructure, reliable data center facilities, and an on-call staff of IT experts to manage everything — support, maintenance, security and upgrades.

The hosted solution model can deliver CRM application functionality without the requirement of the client company to either purchase a software license or to build the necessary infrastructure to run the application(s). It can significantly reduce the need for the client company to hire educated, expensive labor to maintain and administer the system — everything is handled by the HSP for a monthly fee embedded in a service level agreement (SLA). (A SLA is an informal contract between a HSP and its customer that defines the terms of the HSP's responsibility to the customer and the type and extent of remuneration if those responsibilities are not met.)

Hosted solutions are especially cost-effective for CRM technology, which requires a high level of performance, reliability, security and scalability. A variety of business and technical situations make hosted solutions a cost-effective alternative for many within the business community. It can be less expensive than purchasing, deploying and supporting the CRM technology on the business's own servers. But, careful research, due diligence and an intelligent selection process is vital.

While some within the business community understand what a hosted solution provider can deliver and have benefited from the hosted solution model, the community as a whole appears to be unconvinced of the advantages that the hosted solution model could provide. The reasons for such lackluster interest is mainly due to objections by those of the "old school" who have concerns about where software and data are stored, its ownership, and ultimately, how it is safeguarded; and/or concern about the total cost of ownership. Of course, some of this unpopularity may be due to the loss of tax depreciation advantages that accrue on capital equipment purchases.

⊙ ACRONYM ALPHABET SOUP

One of the "duties" of industry analysts and pundits is to invent new acronyms — the more numerous and confusing the happier they seem. It's especially true in the hosted solution field.

Some of the acronyms a CRM team might run across in their search for a hosted solution provider run the gamut from BSP (Business Solution Provider) to SSP (Solution Service Provider), with MSP (Managed Solution Provider) and the tried and true ASP (Application Service Provider) in between. The wordsmiths have even thrown in "ISV" (Independent Self-hosted Vendor) and xSP (generic for any type of solution or service provider) for good measure. Since hosting CRM technology is more of a "solution-based" service rather than a "software-based" service, the author uses the non-specific term "hosted solution provider" or "HSP" to depict CRM in a hosted environment.

The following is an overview of some of the key components that make up the many flavors and subsets of the CRM hosted solution model. This is, in no way, intended to be an all-inclusive rundown of the hosted solution marketplace.

Hosted Solution Provider or HSP (not to be confused with hosted service provider). As far as this book is concerned, there are basically two different business models operating under the HSP umbrella.

The first is the full-service HSP, which run their own data centers. Full-service HSPs do a good job of managing the CRM technology, but must rely on telcos for bandwidth, and vendors for applications, operating systems and hardware. The best offer a large menu of customization options and have a cadre of partners, which gives them the ability to move quickly to serve their client companies. Members within this group compete on not only price and service, but also on architecture, features, functionality, and customization options.

The other business model is the subscription-based HSP. This model offers a standardized package that can be configured but not fundamentally altered, i.e. little customization. This business model defines the lower priced hosted solution market segment. Members of this group compete with each other primarily on service and price. But, I expect the architecture issue will also become *the* differentiating factor as credible CRM suites become available via subscription-based HSPs. This model lets a client company subscribe to their CRM system through a website. Industry numbers indicate that most companies using the subscription model are in the 25- to 30-person user range — clearly small organizations or departments.

Independent Self-hosted Vendors or ISV self-host their software as an alternative means of product distribution. (While some vendors may use a third party to provide hosting, the arrangement is transparent to the end user.) This model doesn't usually offer an aggregated solution. ISVs primarily have expertise in their own applications *only* and partner with systems integrators and consulting firms for the back-end work. Members of this group usually offer their client companies a wide variety of customization options.

Managed Solution Providers or MSPs (not to be confused with a managed security provider or management service provider) hosts the entire CRM infrastructure. They provide a one-to-one delivery model rather than a one-to-many model. They are in essence the next rung on the evolution ladder from the first-generation HSPs. For instance, MSPs offer their client companies integrated solution sets rather than support for individual packages.

MSPs invest their efforts in building the integration layer between applications and resell this capability for a monthly fee. While an HSP may offer complete support for, say, a marketing automation package, a MSP will typically offer to support the package and integrate processes with other applications, such as analytics and call centers.

The MSP helps its client companies to address the technical, financial, and operational issues related to building a highly scalable infrastructure for CRM. Yet, it's difficult to customize in this environment because of the amount of business process re-engineering necessary for customization, thus systems integrators are germane to success. Systems integrators do more than just put applications together, they help with business process and hooking systems up for the solution provider's client companies. Therefore, if using a MSP version of the hosted solution model, check out the integrator partner(s) closely.

Solution Service Providers or SSP (not to be confused with security service provider or storage service provider) add value to packaged software by tailoring solutions. They take a variety of best-of-breed CRM vendor's applications, augment them with a value-added integration layer, and then customize the products for their client companies. This provider is part software vendor, part systems integrator. The majority can deliver an integrated multichannel CRM platform. Again the depth of CRM experience in the systems integrator contingent must be strong for the hosted CRM initiative to be a success.

Wireless ASPs host wireless applications including applications for the CRM market. They may be the answer for a business that depends on field service to keep their customers happy or on a mobile sales force to sell their products and services. This is a good solution for many field service and sales departments that can't afford to put the infrastructure and staff in place to maintain and support a wireless environment with ties to CRM.

As the hosted solution market evolves it will become increasingly difficult to differentiate between HSPs, MSPs and SSPs. Don't go by titles, go by the services and solutions offered when looking at hosted solution alternatives.

✪ WHY USE A HOSTED SOLUTION?

Acquiring CRM software is a major investment for many businesses. In addition, the costs and disruptions that come with CRM implementation can have an adverse affect on the short-term operational efficiency of a company. For example, if the call center is adding field service capabilities to its systems, the field service department must do the same. The marketing department may be gearing up for a marketing automation package but the call center and sales must also be involved. And the IT department supports everything. Everyone must learn how these new systems enable them to interact in ways they previously might not have considered. So, for many companies, the technology end of the CRM process can be highly disruptive AND expensive — it's common for the customization, implementation and training to cost one to two times the software fees and that's before adding in the cost of hardware and networking. (For every dollar a company spends on CRM software, it must spend $2.50 to $5 on integration and imple-mentation.) The hosted solution model can ameliorate some of the costs and lessen the disruption of a CRM initiative.

Advantages that can lead a business to choose a hosted solution model include:
* Rapid deployment.
* Lower start-up and implementation costs.
* Reduction of operating costs.
* Release of capital resulting in a gain in cash infusion.
* Access to world-class capabilities and resources not available in-house.
* Worry-free maintenance and management.
* Improving a function that the company finds difficult to manage.
* The ability to instantly network disparate departments or groups (field serv-ice and R&D, sales and call center, marketing and sales). This is especially important for companies lacking network assets and other systems resources.

Don't let the bad news concerning the financial health of the hosted solution community put a damper on consideration of a hosted solution. Bear in mind that news organizations feed on bad news — it's what grabs the readers' attention. Let me even the score a bit and give the reader some good news about CRM technolo-gy in the hosted environment. It's not uncommon to find among today's HSPs:
* A secure hosted environment.
* Access to CRM technology as if it were in-house.
* Dynamic upgrades as solutions and technology evolve.
* Multiple CRM components (rather than a single application) — sales, mar-keting, call center, web-based self-service, online collaboration, and even mul-tichannel support (telephone, email, fax, web, text and voice chat) — all under one roof.
* Integration of various components with data residing in multiple depart-mental silos (with the help of application integration middleware) for a 360 degree view of the customer.

The Start-Up: For the start-up business or those going through substantial re-engineering (negating the complex integration issues), the HSP approach to CRM technology can work very well. The hosted solution model can put a CRM strate-gy to work without the high up-front investment in hardware and software and without the delay of lengthy development. Little or no software is needed on the end-user's computer. The client company provides a web browser and an Internet connection, the HSP provides the CRM software, implementation services to apply unique business rules, IT support, complete employee training, and main-tenance and management of the system.

The Point Solution: Using a hosted solution for the technology that enables a CRM strategy works particularly well when the CRM project is a point solution (sales force automation) or the focus is departmental (although sharing capabilities need to be in place) such as with CRM in a call center environment.

Outsourcing Divas: Many within the business community have always relied on third-parties for technology and services. For example, it's common to outsource marketing databases, call center services and fulfillment needs. Logic follows that companies within this sector will have few qualms about using a hosted solution for their CRM technology needs.

Costs

Traditionally, the CRM market has been dominated by the enterprise sector. Implementing the technical support for a CRM strategy "in-house" can be extremely expensive. Between hardware, software, consulting, implementation and ongoing support and management, many within the business community find the budget for a full-scale CRM initiative insupportable. The hosted solution model removes the cost barrier. The client company doesn't need to outlay the costs of buying seats of the CRM products since the HSP holds the licenses to the products that they offer. For example, the hosted solution model can offer saving of around 66% on software, communications, capital expenditure, and operating costs.

At the same time the hosted solution model offers key benefits, ease of integration, savings on IT labor, and access to new applications. Other benefits are that LAN islands are now connected to branch offices using HSP/ISP hubs and a connection is available anywhere there's an Internet connect.

Companies with limited means can play in the same ballpark with their more well-heeled competitors.

Implementation

While the benefits of a CRM strategy are substantial, the resources required to implement, manage, and maintain the systems and processes necessary to enable the strategy often stops a business from moving forward — even with a well thought out CRM plan in place. A HSP can make it easier — by implementing the enabling CRM technology via their staff and contract services, and hosting it in their data center. Then they manage and maintain the systems (working in conjunction with the business's IT and other staff) to ensure that the company's CRM needs are continually met. As the business and its needs grow and adapt, so can the hosted services.

Even large- and enterprise-sized businesses, with sophisticated skillsets in their IT department, are finding CRM to be complex and time consuming. As one IT executive put it, "it's hard to do this stuff, it's not just the application or having a data warehouse, it's how to do it properly, pulling everything together in an effective way." Hosted solutions may be the answer. Rather than giving up control, a hosted solution is a chance for a company to gain control over their CRM strategy. Ask an average business how many IT projects come in on time and under budget — IT is a huge devourer of capital.

Limitations

The companies that are in a hurry to roll out a CRM strategy can benefit as much from a hosted solution as the companies with limited development and maintenance resources. Yet, the hosted solution model might not be an ideal choice if there is a need for extensive integration with other systems or heavy customization requirements. Choosing a hosted solution might create an obstacle to a tightly interoperable system that is customized to a company's specific requirements.

This is where a careful research and an intelligent selection process comes into play. The hosted solution model can still make sense — just know the limitations before signing on the dotted line.

A Multiple HSP Platform

As HSPs gain popularity, some client companies will inevitably demand best-of-breed solutions from multiple HSPs, engendering the problem of how to efficiently share data among HSPs and at the same time deliver an integrated application view. So far this issue has received little attention, but it could well be a defining issue in the future.

Advantages

A hosted solution provides laudable cost metrics, an accelerated deployment speed, and skilled support resources. Unlike traditional client-server CRM applications, there is virtually no software to own. Consequently, there can be a substantial reduction in software management tasks and related costs.

Again, the CRM application(s) is accessed over the web with the HSP maintaining (hosting) both the application and the database for a monthly per-user fee. This can reduce both software and system administration costs while providing clear cost of ownership metrics. With the hosted solution model, there is a substantial reduction in the many tasks associated with maintaining traditional client-server based CRM applications. Generally, the HSP handles all system maintenance tasks, including back-up and recovery, upgrades, network management and user additions and deletions.

✪ DIFFERENCES BETWEEN HOSTED SOLUTION MODELS

The hosted solution providers are nothing if not accommodating — there are an amazing variety of players in this fast-evolving niche. But, all HSPs aren't vying for the same customer base.

Application Specific

Some HSPs just host one or two CRM vendors' applications with little, if any, customization of the software. Their mission is to concentrate almost entirely on supporting the application. This group will have significant service strength related to their specific CRM offering(s) and can accelerate implementation schedules.

If the CRM strategy is to begin with a point or modular approach, the gut reaction of the CRM team may be to go with a HSP that offers an exceptional point solution, not an HSP that offers technology that can enable a complete CRM strategy. The single-application focus is a benefit to companies looking for the most qualified people to implement, maintain and manage a key application, but it can be a drawback if a company wants to grow their CRM initiative through additional functionality that is only offered by other vendors.

In the long-term the company may find itself dealing with a disparate cadre of providers that don't work all that well together. Yet, if only implementing a point solution or pilot project, this flavor of HSP is a good idea.

Tailored Solution

Hosted solution providers can add value to packaged software by tailoring solutions. For example, they can take a variety of best-of-breed applications that work in a hosted environment and augment them with a value-added integration layer; for instance, an agent desktop and give it a chat functionality and CTI to create an application that can be hosted. Many hosted solution providers then take the results and integrate and prototype it to decrease the time to market. Now, when a business comes knocking on their door looking for a specific hosted solution, the HSP can easily customize its product offerings for that business's specific needs. This model is a good option for a business that needs a CRM project up and running in record time.

Vendor Specific

Many HSPs align with a specific CRM software vendor. The vendor supplies their full product range of CRM applications (or suite of products) and the HSP manages them, providing implementation and integration, data center management, connectivity and support. Others are actually CRM vendors (the ISVs) who use the hosted solution model as a new distribution model for their own applications, with all the necessary accouterments — hardware, infrastructure, network, security, data center facilities, and IT staff.

Traditional CRM vendors who host versions of their solutions will eventually represent a sizable piece of the hosted solution market share. Although many CRM vendors won't put themselves in the HSP market space, they do meet the HSP definition — they offer software for rent over a network.

With the proper research and due diligence, this HSP model might be just the answer for a small to mid-sized business that wants to implement a corporate-wide CRM initiative.

Menu of Solutions

Some HSPs try to be all things to all comers. While such HSPs offer a menu of solutions from various vendors, their services aren't all inclusive. They don't normally provide, for example, the building and/or maintenance of integrated system architectures. In most implementations, this group of providers performs little, if any, customization of the software. Instead the provider concentrates almost entirely on supporting the application.

While this model may be a bit troublesome to work with, it can provide a business with CRM solutions that "fit" a specific need on a project by project basis. This allows a business with a corporate-wide CRM initiative to avoid the necessity of managing a disparate group of HSPs.

Vertical Solutions

There's also the HSPs that cater to a specific vertical market. Different vertical industries interact with their customers in different ways (the financial industry differs greatly from the manufacturing sector and both have little in common with the health care industry). This group of HSPs usually provides highly customized CRM technology. The CRM team just needs to perform the proper due diligence. These HSPs should possess specific vertical market expertise. What these HSPs offer regarding integration support and customization can vary widely from provider to provider.

This is another hosted solution model that can enable a business to rapidly launch a CRM initiative and obtain a fast time-to-benefit.

End-to-End Solutions

The *créme-de-la-créme* of hosted solutions goes beyond hosting and maintaining applications. They provide extensive customization for their client companies. In this group are full-service HSPs that have developed strategic partnerships with vendors, integrators, and consultants to provide extensive customization and integration support.

These HSPs offer an end-to-end solution (whether a single vendor suite or a set of applications from a diverse group of vendors) that can feed into other systems, such as ERP or accounting applications, as well as offering some management reporting tools. There are also HSPs within this category that offer suites of best-of-breed applications that run together on one platform. Many of these best-of-breed suites offer a depth of functionality and business process expertise.

The cost will be greater and the implementation period and time-to-benefit will be longer than with the other hosted solution models. However, the end results will be more akin to an in-house system.

Business Solution Providers

Business Solution Providers can provide the whole cow. These solution providers identify what products and partners they want to work with and then invest the time and the infrastructure to integrate all of those applications before ever offering any solution to the business community. (Some BSPs are even targeting vertical markets.)

Many are both integrator and an aggregator, meaning that they can provide an integrated suite of applications or they can aggregate specific applications to a third party. For example, it's possible to find a BSP that provides CRM tools, budgeting tools, accounting tools and order processing tools, i.e. core services in almost every area of business from front-office to back-end. Once the CRM team has made their selection from the menu of applications offered by the BSP, the BSP becomes essentially a consulting company to integrate everything. When a solution is crafted and purchased the applications they offer already "talk" to each other. This sets BSPs apart from most HSPs. Every application is pre-integrated with the others — all the works was done beforehand. The implementation stage is essentially only to get the company's staff up and running on the chosen applications and then to import the necessary corporate data into the hosted systems. Therefore, since the BSPs typically offer applications that are pre-integrated, they *really can* get a client company up and running within 60 days.

⊕ BENEFITS

The more pressure a business and its CRM team are under to launch a CRM initiative, the more willing all may be to go with a hosted solution model. Also, many businesses appreciate the predictable monthly costs and the flexibility they gain in terms of managing the growth and complexity of the applications.

There are big benefits of the hosted solution model, especially for small to mid-sized companies — less up front cost, quicker time-to-benefit, access to hard-to-find IT skillsets, and even consulting and integration services.

Time-to-Benefit

Time-to-benefit or the elapsed time between the start and the end of the implementation of a CRM solution is becoming increasingly a key selection criteria for companies intending to implement a CRM strategy. Some companies want a quick fix, and hosted solutions represent one. Time-to-benefit is a compelling reason to consider hosting a company's CRM technology. Many CRM teams can't provide a quick implementation timeline for a CRM initiative even when they bring onboard a truckload of technical partners. When their executive board comes to them and says, "We want CRM and we want it now," the savvy CRM team will take a hard look at the hosted solution model.

When it comes to time-to-benefit, the hosted solution model generally wins hands down. A hosted solution-based CRM implementation compresses the delivery time. Many CRM hosted solutions are typically up and running in as little as sixty days (but can take up to four months). By contrast, a traditional in-house rollout of a typical CRM package could take several months to more than a year.

IT Skillsets

Beyond time-to-benefit, early adopters of hosted applications cite the challenge of maintaining up-to-date consulting and IT skillsets as an impetus for selecting a hosted solution model. Finding skilled people to run and maintain CRM applications is less a problem then keeping them current in the torrent of new skills required to do their jobs. I'm talking about database management, C++, web, XML and HTML development skills, communications skillsets and so on. For many IT departments, these are disparate skills that go way beyond their traditional hiring practices. For them it's more practical to find a HSP that specializes in those areas.

The Sweet Spot

When experts talk about hosted CRM, they often cite the benefits for midsize companies, and with good reason — that market is definitely a sweet spot. Midsize companies typically have good, reliable Internet connections coupled with budgets that make hosting a sensible option. For those just now rolling out a CRM initiative, it's advisable for them to consider the hosted solution model.

But, what about larger corporations? Most analysts give thumbs down to HSPs for the large to enterprise market, and their reasoning makes sense. Among other things, this market usually has the resources and experience to keep their CRM applications in-house. They also have ample budgets, which will allow for large-scale implementations. In many cases, enterprise IT departments have more experience than any HSP could possibly have on their payroll. Enterprise-sized businesses have also often already made significant investments in in-house CRM applications.

Still, this market isn't shunning the hosted solution model wholesale. While this group wouldn't abandon the CRM applications that are already in-house, they may consider the hosted solution model for new applications. Many times, if a new CRM solution lends itself to being hosted, and the time-to-benefit exceeds the IT shop's ability to bring it in-house or requires certain skillsets that the department is having difficulty finding, they will shop around the HSP venue.

Web

Typically, applications provided by hosted solution providers are written in HTML, XML and Java, and delivered to the client company through a web browser. The web, as a means of distribution, totally eliminates one of the long-standing headaches of IT: getting the entire company to standardize on a single platform. Giving a company's employees remote access can be as simple as entering an IP address in a dialog box. In fact, the ability to access a CRM application remotely is one of the key reasons for going with a hosted solution.

✪ THE CONCERNS

What issues must HSPs address as they take their new and improved models to their chosen markets? Well, there are a few items that can make companies think hard about whether they really want to go with hosted CRM applications.

Financial Health: One of the main issues that cause many business to hesitate using a hosted solution is the financial health of the HSP. According to industry analysts, over half of the HSPs in business at the end of 2001 will not be around by the end of 2002.

Security Issues: The other primary concern focuses on security within the hosted solution environment. More than one-fourth of HSP's customers will experience a security incident at the HSP that will result in the compromise of sensitive data.

Vanilla Software Packages: Every company states that the fact that most HSPs only offer vanilla packages is a concern. Companies don't want a vanilla package when it comes to CRM. Yet, vanilla packages are what most HSPs offer. Only a few HSPs offer more than minimal customization. The typical HSP shies away from customization either because they lack the necessary programming talent, or because customization inhibits their ability to perform mass upgrades.

Customization and Integration: Although listed last, customization and integration issues are among the top five concerns of most IT departments when considering a hosted solution for CRM systems. Most companies using a hosted solution have had a variety of problems related to customization and integration, or more specifically, to a lack thereof. CRM tools are some of the most customized and integrated applications that a company has in its technology portfolio.

Although, at the moment, most HSPs find customization a challenge that they are unwilling to undertake, as the hosted solution model evolves, more and more providers will offer credible customization and integration capabilities.

✪ TRADE-OFFS

Depending upon the point of view, many of the benefits that are derived from a hosted solution can also be viewed as trade-offs. Four related and somewhat obvious trade-offs are control, access, dependence and ownership. Let's consider each one separately.

Control: Clearly, giving control to a HSP is a doubled-edged sword. The client company must trust the HSP to follow good systems procedures for disaster recovery, user support, system upgrades, bug resolution, version control, and data security. At the same time, the client company will experience some loss of control and less management oversight.

Access: Though Internet connections are very stable, a company runs the risk of losing access to the CRM application(s) in the event the connection to the HSP is lost, unless the web-based application can operate in a "disconnected mode" (not connected to the Internet or Intranet). This raises concerns about network reliability, privacy, uptime and connection speeds.

Dependence: Because the server is hosted, the company is dependent upon the HSP to keep it operational. It's noted, however, that world class data centers are significantly more redundant, secure and better supported than most in-house systems. Just choose the right HSP.

Ownership: Depending on how the contract is structured, the company may not own the actual CRM application (this is one of the ways that cost is reduced). For many companies, avoiding the cost of the application is actually one of the greatest benefits of the hosted solution model. (In all cases, however, the company will own its data.)

As the hosted solution model matures, client companies' confidence and comfort levels will follow. In the meantime, ask the HSP to clearly document how it intends to address these concerns. In particular, what guarantees can the HSP provide that the company's data will be secure and that the application will perform reliably over the Internet?

⬡ THE DECISION PROCESS

To help in the decision process, the CRM team should determine:

* What are the system requirements of the proposed CRM initiative?
* What is the overall impact of a hosted solution on the CRM strategy?
* What functionality does the company want and need in the way of CRM technology?
* What is a realistic implementation timeframe for having the CRM initiative up and running?
* What is the potential cost saving if the company signs on with a HSP to provide the enabling technology for the CRM initiative?
* What are the expectations within the company for the CRM initiative?

Due Diligence

Paying someone to host the technology to enable the company's CRM initiative might save time and money, but be careful when choosing the HSP, just as with any other strategic partner. Along with looking at a hosted solution's functionality and agility, perform proper due diligence. For instance examine the potential HSP's:

* Certification (such as a vendor certified outsourcing partner).
* Applications. Are the hosted solutions built upon best-of-breed processes?
* Field capability. Is there a good record?
* Relationships with software vendors, ISPs, and implementation and infrastructure partners.
* Proven history.
* Experience with the company's system(s).
* The staff and management. Is there an overall good experience rating and in-depth business process knowledge?
* Reputation with industry analysts.
* Client references.
* IT infrastructure. Does the HSP own it?

* Delivery method. Does the HSP have a reliable high-speed connection and a robust networking fabric that ensures dependable communication between the company and the HSP?
* Customer care. Are there 24 x 7 access, a high degree of technical skill, adequate staff, etc.?
* Security procedures. Is it best practice? Can the HSP, for example, protect the company's data from unauthorized access?
* Financial viability.
* Contract terms. How flexible is the contract? What are the termination procedures that will enable the company to get its data back?
* Culture. Is the HSPs culture a good match with the company?
* Guaranteed service levels. Does the HSP have an infrastructure that can ensure the SLA.
* SLA. Take a close look at the SLA in order to make sure the provider is held responsible for things such as security lapses.
* Technology cycle. What is the HSP's technology cycle? One of the key benefits of the hosted model is that it keeps customers on the cutting-edge of technology without the traditional associated challenges. However, find out how often the company will need to change out solutions. Ask what the cost will be to put it together and what will be the ramifications on both business process and outbound process.
* Training. Does the HSP offer onsite and online training for the client company's staff? Is it a one-shot deal or ongoing? (Often only about 10% of any CRM solution's potential is ever used because of poor training.)

On a business level, find a HSP that:

* Will zero in on solutions for furthering the company's CRM goals and objectives.
* Won't wedge the company into a proprietary solution. If the HSP delivers proprietary pieces from its data centers on up, it isn't offering a best-of-breed solution (no matter what the marketing material states).
* Offers a contingency plan in the event of a software vendor's bankruptcy, acquisition by a competitor, or other problems.

If the CRM team has chosen to go with an HSP that offers tailored CRM applications, be sure the HSP can provide:

* An integrated knowledge engine to drive the necessary processes with the company's business logic and rules.
* A proven telephony routing and Internet-based communication.
* Internet chat, collaboration, email and web call-back communication, if required.
* Technology and customer service experts to design and deliver not only implementation services, but also ongoing performance monitoring and improvement.

* Platform customization to match application functions to the company's needs. Also find out:
* What are the options available for integration with legacy systems?
* How is the CRM system administered (adding/deleting users, adding/deleting fields and records, bug fixes, installation upgrades, etc.)?

Security

Most HSP operations are an invitation for hacker activity because by their very nature HSP's systems have to be open for the client companies to get into them, causing a real challenge in safeguarding them. But also, many use hacker vulnerable out-of-the-box Windows and Unix configurations. Then there's the "phantom computer dilemma." This occurs when a HSP adds a computer to its network for a dedicated project, but leaves the machine on the network once the project is finished (a typical scenario). Since only a few people are apt to know of the computer's existence, it will probably be unsecure and thus, become a target for hackers.

Since most HSPs and client companies assume that the data will physically reside with them, the issue of security must be stringently addressed. The CRM team should ask what the HSP's security model is and what processes and technologies have been put in place to implement them.

Note: *The engineering necessary to quell the "ownership of data" objection is "old hat" so the data ownership issue is a non-issue. To state that a client company's data will be protected and portable when and/if the company chooses to retrieve and move is just a marketing ploy.*

Carefully research the HSP's security procedures by asking the hard questions, such as:

* How is the client company's data kept secure and protected from unauthorized use and unauthorized users?
* How often is the data backed up, what are the backup methods and has the HSP recently successfully restored data from a back-up source?
* What is the HSP's security and acceptable use policies? Ask to review the written documentation. If the HSP refuses, it's probably an indication that it doesn't have formal security policies — a definite cause for concern.
* Does the HSP have a formal emergency response team? What are their qualifications?
* Are the latest security patches applied to all systems immediately?
* How often is a security audit performed?
* Does the HSP have a security specialist on staff? Under contract? If so, ask to see the specialist's credentials.
* Does the HSP keep an ongoing record of vulnerabilities that have been addressed?

✳ Does the HSP have multiple layers of security? Security concerns are easily addressed through the use of encryption and Virtual Private Networks (VPNs). VPNs make use of the public telecommunications infrastructure, maintaining privacy through the use of a tunneling protocol and additional security procedures. Data is encrypted before it is sent through the public network and decrypted as it is received. Make certain that the service provider doesn't rely on either password or cryptography technologies alone.

✳ If the company operates within a vertical industry, such as medical or financial services, which have specific legal or regulatory security requirements, determine if the HSP is equipped to conform to the necessary requirements.

Note: Corporate data is at risk whether it's sitting behind the company's firewall or a hosted solution provider's firewall. A company's data is likely to be more secure in a hosted environment than in-house, because unlike the typical corporate IT architecture, security is a core requirement of any viable HSP.

Stability and Reliability
Next, are questions concerning the HSP's stability and reliability.

✳ Is uptime guaranteed?

✳ What procedures are in place to support the HSP's claim of reliability (fault tolerance, system redundancy, etc.)?

✳ What are the systems infrastructure and business processes in place to ensure reliability during any high-growth phase the HSP might experience?

✳ What is the technical and application support? Is it adequate?

✳ What are the disaster contingency plans in case of fire, power loss, or other catastrophic events?

In short, look for comprehensive SLAs, documented disaster recovery procedures, data security statements and descriptions, scalability benchmarks and proven technical support. Just as important, ensure the HSP under consideration is financially fit so that the company's data will not be held hostage in the event of a bankruptcy or other catastrophic failure.

Don't forget references. Get at least five (ten is better) references. Then contact each reference; quiz them on their experiences with the HSP.

If the company enters into a relationship with a HSP without performing a thorough due diligence procedure, the company will, in all probability, fail to reap the full benefit that can come from using a hosted solution.

⊙ INTEGRATION WOES
Still, hosted solutions aren't CRM's Eden. The expense of a software license is only a small part of the costs of implementing a major CRM system. Industry observers agree that integration presents a huge challenge for users of hosted CRM. Hard wiring and integrating a hosted solution provider's offerings with existing in-

house applications isn't always an easy task. The back-office is where substantial amounts of critical information are stored. For CRM to work, integration to back-office systems, especially ERP, is critical. Many hosted solution models can't, at this time, provide strong integration to these systems.

Time

Time to deployment is a key factor in opting for a hosted solution. Ultimately, the main driver in a company choosing a hosted solution is that hosting puts companies on a much faster track to an up and running CRM initiative. Many within the hosted solution community tout a 60 day or less time-to-benefit. Admittedly, a simple hosting setup can be accomplished in a scant 60 days, but the CRM team should document what is actually delivered in 60 days — many HSPs place a limit on the functionality they will provide in a shortened time frame.

⊙ COSTS

Leading industry analysts estimate that going the hosted solution route can mean a 30% to 60% cost savings over the traditional implementation model, but a hosted solution model should never be viewed as a financial Utopia. In fact, moving to a hosted solution model might not provide much cost savings. The savings really depend on individual cases.

Many executive boards will ask about the total cost of ownership (TCO) of hosted CRM solutions compared to traditional CRM licenses. Any TCO calculation is, by itself, an incomplete measurement since it can only determine in monetary terms which product or service is more or less expensive. Conspicuously absent from such calculations is any discussion of intangibles such as overall convenience or the peace of mind that comes with one solution versus another.

Companies that adopt a hosted solution model on cost alone are missing the point. What matters is how quickly the company can realize an economic benefit, i.e. time-to-benefit. If the system can be up and running in 60 to 90 days with a prepackaged solution, rather than taking that time to assess technologies and gather systems integrators, then that is a valuable proposition.

The CRM team must consider the value beyond the base cost. Look at the hosted solution from a TCO perspective, but also factor in the soft costs of downtime, the ability of the IT staff to better predict and respond to trends corporate-wide, and impact on the customers. If the quality and quantity of what's delivered is also considered, it's possible to see the impact that the hosted solution route can have on a business's bottom line.

All-in-all, comparing the overall costs of implementing a CRM package in-house and outsourcing the job is a tricky endeavor. ROI calculations are complex because most companies never really know what internal systems are costing them. Rather, the key comparison points are the hassle factor and deployment time frame.

The hosted solution model is still immature, thus actual fee structures are all over the map. There are HSPs who charge $29.95 or $50 per month, per user and others who are charging a flat $5,000 fee for an entry-level contract and $50,000 per month or more.

Yet, when hard numbers are put to the hosted solution model, i.e. weighing the cost of renting against purchasing and then having to capitalize everything, for many companies a hosted solution is indeed more cost effective.

According to GartnerGroup, 50% of midsize corporations will host some portion of their applications, and of those being hosted, 10% will be CRM applications. Furthermore, 40% of large corporations will host business applications, and of that 40%, 5% will be CRM applications.

GartnerGroup also reports that client companies leasing an application from an HSP can save between 33% and 53% over purchasing and managing the hardware and software themselves.

✪ IN SUMMARY

The hosted solution model doesn't just provide a sustainable competitive advantage, but a renewable one. In the right circumstances, a company can greatly benefit under this model, and the interest level is growing.

Making an exact comparison between the various hosted solutions providers is difficult because of differing functionality, service, and potential for customization. Many companies may find that the ability to customize is key to their ability to succeed with a hosted CRM application. On the other hand, other companies will find that the vast majority of changes to a CRM system can take place through configuration options instead of wholesale customization. The right answer will depend on the exact business problem to be solved.

It's a common feature of high-tech life that a good idea can be re-arranged and significantly improved by the forces of market demand, continuing innovation, and entrepreneurship. CRM market forces, such as client security and total cost of ownership concerns are among the foremost influences re-working the basic hosted solution model, which should turn it into something that will be of optimal utility to the vast mainstream market.

There is a hosted solution for virtually every CRM need. There is also little doubt, with the wide and diverse requirements of the marketplace, that the best in each area will thrive.

Chapter 15
Partnering for Success

The problems that arise with a CRM initiative almost always can be attributed to inadequately addressing the complexities within the CRM strategy itself. A CRM initiative is a complex undertaking; due, in part, to a lack of inter-departmental coordination abilities, deficient change management capabilities, and a dearth of necessary skillsets for integration with existing IT infrastructure. It, therefore, may be necessary for the CRM team to partner with not only CRM vendors, but also with consultants and systems integrators to provide adequate technical support for the CRM initiative.

The first step in determining what partnerships are necessary is for the CRM team to define the risks, the mitigants, and the level of executive commitment. Next, the team must develop a road map for implementation and the value metrics for each step. This allows the CRM team to spell out delivery commitments, as well as, to measure and communicate goals and achievements.

Hiring a tech partner is like bringing in a trusted advisor or guide. Pick one who's familiar with the terrain, if the company is a manufacturer with legacy systems, find a partner with CRM experience in that environment. Choose one whom everyone working on the initiative feels some chemistry. Once onboard, the tech partner should be considered a key resource.

If the company has an existing working relationship with a tech partner outside the CRM initiative, then they may be the obvious choice to work with on the CRM project. However, don't saddle the CRM initiative with tech generalists.

Aside from competency to work with the company's specific CRM initiative, the next most critical thing to look for is a cultural match. Most

CRM-related meetings are long and detailed. Everyone should feel comfortable and able to express themselves. If not, these meetings can be unpleasant and unproductive.

⊙ HOW TO FIND A TECH PARTNER?

If the company doesn't already have a working relationship with a consulting or systems integrator firm, how can the CRM team find the right technical assistance? While there are no central resource that provides in-depth information on CRM consultancy or systems integrators who are well versed in CRM implementation, there are resources a CRM team can mine.

* Read industry literature.
* Attend technical conferences and trade shows.
* Online communities can be helpful and informative.
* CRM vendors — even if the CRM initiative isn't at the technology selection stage, the CRM team can still contact CRM vendors and ask which consultants and integrators they're familiar with.
* Attend consultants' or systems integrators' presentations (also read their journal articles) to evaluate how much they really know.
* Customers and supply chain members. Talk to them and ask whether they've instituted a CRM project. If so, ask which tech partners they use or have used in the past.
* Re-visit tech partners the company has used in the past. Assess these firm's CRM capabilities and limitations before deciding whether to continue the search.
* Obtain referrals from other similar positioned businesses, trade associations, or referral services.

⊙ CONSULTANTS

The first outside partner a CRM team typically considers is a consultant. But, for every successful CRM initiative that benefited from an outside consultant, there's another with a tale of woe. It pays to research the exact needs for a consultant before signing any consulting agreement.

While consultants might help to improve the success quotient of the CRM initiative, the CRM team must not forget that the company's executives, management and staff are the experts in the operation of the business and interaction with its customers. The consultant should *enhance* that expertise, *not be a substitute* for it. Also, the CRM team should not depend on a consultant for decision-making — including purchasing decisions. It's not uncommon for consultants to receive a commission when the company (as their client) buys a product or hires someone the consultant recommends.

Not all consultants do all kinds of CRM-related work, and not all of them can perform every job equally well. Before hiring a consultant, identify that consultant's expertise, qualifications, professional certifications and interest in the company's CRM initiative. Then determine if there is a fit with the needs at hand.

Note: Bringing a consultant onboard to help in the technology selection stage typically costs between $10,000 to $40,000.

Why Hire a Consultant?

At one end of the consultant conundrum is the "let's do it ourselves" group — damn the torpedoes, full speed ahead.

A little further down the continuum is the "we only use consultants, a little bit (through our CRM software vendor)" contingent. These folks have their head in the sand. The companies that rely solely on software vendors for the necessary consulting services are treading on thin ice. While some CRM software vendors discourage their customers from buying a product that's not a fit, most vendors just want a sale, they worry about fitting the client company's needs around their product's capabilities later. That's not what I call consulting. The outcome is bass-ackwards —starting with technology and then trying to back into re-engineered work processes, then redefining roles and responsibilities, then developing customer-centric strategies. What a futile exercise!

At the other end of the line is the "our CRM initiative relies on consultants" segment — this group suffers from the Experts Must Do It Better Syndrome.

But where is the middle ground? Why not limit the use of consultants to what companies can't do or learn on their own? Consultants can provide objectivity when an outside perspective is needed. They can teach internal staff specific skills when needed. They can help the implementers stay focused and organized. Some can help select the right software. And good ones can point out the pitfalls.

When the in-house staff lacks in-depth CRM experience, a consultant can be of help, particularly when the company begins to meld sales, marketing and service functions together. If the CRM team is ineffective, tentative or non-existent a consultant may be a good idea. Another situation where a consultant can be of benefit is when top management buy-in for the CRM initiative is shaky or absent. Demonstrating defined goals and objectives of a CRM initiative is crucial in obtaining management support — a consultant can often be of help during this process.

There are five broad operations are typically where a CRM consultant may be of value.

Adoption of customer-centric values (without which, there's no CRM): Can CRM consultants help here? Sometimes, but only to put "the fear of the market" into the top brass. Although sometimes a consultant can help in the change management process, he or she must be experienced in working with executive board-types in order to be the "moral compass" the executive board needs. To tip the scales, look toward a specialist — a change management consultant.

Still, even a change management consultant can't accomplish much unless the head honchos are open to change.

Another area within this sector where a consultant can earn their pay is in the planning methodology used for the adoption of a customer-centric business model. It's so different from conventional planning that a good, strategic CRM consultant trained in the strategy discipline can contribute by guiding the planning process. Also, the right consultant, at this stage can provide an objective voice that fairly reflects the will of the customer.

The Re-design of Workflow: New work processes must be driven by new workflow that arise from the new customer-centric strategies that come with a customer-centric business model — it's the only way to provide the proper impetus for change. Although workflow re-design decisions should be made internally, experienced CRM consultants know techniques for workflow mapping that can bubble up issues that the CRM team might not have seen. Once these issues see the light of day, the company can make the necessary decisions (which will affect departmental roles and responsibilities). So CRM consultants can be of help here. Especially if they're expressing, through new processes, internal decisions already made.

While re-defining internal roles and responsibilities are necessary for a customer-centric business model to work, it isn't the role of a CRM consultant to change how sales, marketing, service, accounting, and/or manufacturing employees do their jobs — that's the task of corporate management. However, a CRM consultant who is trained in the business process redesign discipline and skilled in workflow rather than technology (very seldom is a consultant good at both) can make *recommendations*. On the other hand, if a company's concern is to get help implementing its reorganization decisions *after* it makes them (most large companies do need help), then go for a consultant experienced with the strategy discipline, such as a change management consultant.

Process Re-engineering: Initial process re-engineering, i.e. getting the process straight, then automating it is an idea that has become axiomatic in the CRM field. Automating broken processes only creates more problems and results in wholesale dissatisfaction with the CRM project. This work is so thankless that many will job it out just to get rid of it — a business process redesign consultant with mapping automation software can turn weeks of work into just a few days.

Technology and Implementation: This sector includes deciding which software application best addresses the CRM strategy's needs and then implementing the technology. This can be a good job for a consultant with technical implementation experience. Gathering data and company-wide agreement for a Request for Proposal (RFP) can consume the time and efforts of an internal staff. Most are already so busy with their own departmental demands and schedules that it's difficult for them to give the necessary attention to the selection and implementation process.

An objective, neutral CRM consultant well versed in the technical implementation discipline can help to organize the CRM initiative — cobbling technology requirements into a cogent requirements definition document, compiling a "long list" of vendors that meet general criteria, helping to cut the list down to a short list, then issuing an exacting RFP. But remember that the CRM team should make the final choice (but only after it has received and parsed significant system user input).

Note: A deep computer-science background, experience with specific programming languages or even experience with broad application areas such as PeopleSoft, Infomix or Oracle is just the beginning of the skillset a consultant should bring to a CRM initiative. Most CRM consultants need, for example, a strong database architecture background and possess competent analytical skills.

Consultants can help bridge the gap between what the company really needs and what everyone thinks is wanted and needed. As one executive with a CRM initiative that had to be scrapped and started over stated, "The first time we went out and bought technology and then a consultant, we should have done it the other way around."

CAVEAT: Keep in mind that most consultants already know what technology they will recommend — it's what they are familiar and comfortable with.

If it's determined a consultant can bring value to a CRM initiative, find one that can operate in a three-dimensional world — someone who can think through various technology options and understand their business effects. CRM requires skillsets that can mix together various business processes and systems. The team should look for a consultant that is part sales person, part technologist and part architect — attributes not usually found within the corporate IT department.

What to Avoid?

Implementing CRM can entail a lot of painful lessons if everyone doesn't know where the pitfalls are. Some of the problems are obvious. There are consulting firms, dazzled by dollar signs, that think the fastest way to cash in is by cozying up with software vendors; trading their objectivity for a piece of the software action.

Keep the IT department in the loop. Before hiring a consultant make absolutely certain that the IT personnel are in agreement about both the need for the CRM initiative and the choice of the consultant. IT personnel often feel uncomfortable with consultants that other, less technical, department personnel hire. So, don't do it — the IT department is too important to the CRM initiative. If the consultant is willing to sign the contract without IT's approval, don't hire that consultant — keep looking.

When a consultant is engaged to perform a requirements definition, draft the RFP or help in the system selection, avoid buying that recommendation and the consultant's services to implement it *based solely* on the consultant's say so. The flaw in this type of arrangement is that the company has fundamentally given up the right to know what it is buying. To give a CRM initiative a good chance for success, the company's staff must be involved every step of the way. This ensures that the technology purchased best *meets, not defines*, the company's requirements.

Don't go running to consultants with the idea that they will tell the company how to run its CRM initiative — the company just might end up with "bolted-on" CRM applications that fall apart just about the time the consultant cashes the last check. Don't allow a consultant to deliver someone else's best practices so the company doesn't have to develop its own. The CRM team must participate every step of the way.

Consultants Wrap-up

What is not the consultant's responsibility? The entire CRM initiative from strategy through implementation — they can't design it and deliver it in on a silver platter. A consultant (or even a battalion of consultants) cannot substitute for internal participation, and they absolutely can't implant values that must grow from within the corporate walls.

A consultant can bring an objective viewpoint — sometimes a CRM team is too close to objectively define what is needed. Some consultants are also good in dealing with internal political situations that might have a CRM initiative mired down.

Bear in mind, consultants aren't super heroes, they can't fix every problem or make everything happen. In the end, a successful CRM initiative is in the hands of the company, its executive board, management and CRM team.

⊙ SYSTEMS INTEGRATORS

When a customer contacts a business, most of the time they are looking for a product or a service or information or advice before beginning the purchase process. The more complex the product or service, the longer and more detailed the questions. To enable employees to answer these questions, the company either provides the appropriate staff with high-level training or it refers the customer to an in-house expert. A similar process occurs when a company decides to add new systems, enhance its existing systems, or to adapt its existing systems to enable a CRM strategy.

The CRM team may not possess the in-house expertise needed to assess its requirements and its IT department may not have the time or resources to assist the CRM team. In this situation, a company may seek out a consultant for assistance or sometimes it relies on the vendor community to provide the expertise. But other times an independent, experienced and professional systems integrator is necessary to determine the technical needs of the CRM strategy. An independent systems integrator can help to select, install, test and support products to meet the CRM initiative's technical needs.

Most point solutions don't require the services of an independent integrator. But, if the product is very complex, or if there are a number of point solutions to be melded together, or if there are archaic or other troublesome legacy systems, then hiring an independent systems integrator is, most likely, a good idea.

For instance, a business with a large number of direct customer relationships will typically have multiple customer-facing departments, and each department will, most likely, have independently selected its own CRM system. Integrating all these to back-end systems, and keeping any new changes in sync, is a continual nightmare for internal IT departments. Systems integrators can come to the rescue.

Systems integrators can function somewhat like a personal shopper. In addition to making sure that the new CRM systems can work well by themselves, the integrator can also verify whether the new systems are compatible with other components within the corporate technical architecture.

Just as a new suit often has to be tailored before it fits correctly, so it is with CRM technology. The integration aspect of systems implementation is still a necessity despite the advent of products built to open standards and despite the emergence of more reliable components that make integration easier.

Open standards don't eliminate the need for systems integration. There's not a single CRM product that is truly ready-to-run, i.e. every CRM application needs some kind of customization.

A systems integrator (whether hired independently by the company or supplied via the vendor) can provide a single source of support. But, not all systems integrators are competent in all kinds of CRM-related work. Identify their expertise, qualifications and interest in the company's CRM initiative. Request their marketing materials, including relevant educational background, experience and professional certifications.

Ask prospects for specifics (i.e. do they have the expertise for the job at hand) and ask for references. The referenced companies' IT ecosystems should be similar to the company's.

As with the products they work with, each systems integrator has unique strengths and scope. There are integrators that design their integration efforts around, for example, the call center environment, i.e. open, switch-independent systems that work seamlessly with other components such as IVR and workforce management systems. Other integrators have a staff of programmers and project managers who can install and can develop software or middleware to help a company implement specific components — this should be done only when there's no other options. Yet, custom-built components are often necessary in the call center environment due to the numerous proprietary products usually found within that architecture.

Not every CRM initiative requires the same level of integration, and systems integrators offer different tiers of services. A few systems integrators also offer their own products (especially within the call center arena), which is different from custom-built products. For example, these can include call transfers with screen pops, skills-based routing, IVR scripting, multimedia queuing and web callback.

Systems integrators apply an increasing amount of structure and discipline to a CRM project. To save time, systems integrators may reuse software or middleware that they have previously written, such as interfaces between IVR systems and host systems. While such reuse can save two to six weeks of effort, many times it's better in the long run to use off-the-shelf components.

One of the most important services some systems integrators can provide is showing a staff how to use CRM software to identify and segment the company's best customers. They can examine every detail of the systems that the company uses to communicate with its customers. Then the integrator applies a methodology to ensure the systems recommended allow the company to provide consistent service regardless of how customers choose to reach the company.

How to Find the Best Fit?

The CRM team doesn't need to worry about being trampled by a stampede of integrators screaming for the work. *Good CRM integrators* are still rare. Basically the CRM team will be among the begging masses at the door of the systems integrator rather than vice versa.

What should a CRM team do when considering the possibility of hiring a systems integrator? Formalize an effective selection process. Then write a summary of the CRM initiative including its objectives and technical needs. This summary can be a handy reference tool when making initial contact with perspective integrators.

The real trick is finding a systems integrator that is intimately familiar with the company's technical architecture. Three factors should be considered: specific application experience, specific legacy system experience, and personalized service.

To obtain the greatest benefit after the installation of new CRM strategy and technology, the CRM team should seek systems integrators that will transfer much of their knowledge to the company's IT department. The process of knowledge transfer should be a key part of any systems integrator's contract.

When seeking out a systems integrator find a firm that will provide a single point of contact for the entire integration project and for support after the project is done.

Note: *Many independent systems integrators offer a maintenance agreement whereby the integrator regularly checks the installed CRM systems to correct potential problems before they occur.*

But the systems integrator's methodology must also be considered. To cope with shorter time frames, some systems integrators prefer to implement projects in sections rather than all at once. A phased implementation methodology has many advantages. It enables the company to realize the benefits of a project sooner and receive immediate feedback about specific steps during the project.

Systems integrators sometimes segment the phases of a project even further. For example, the integrator sometimes chooses to set up a new SFA and add the wireless connectivity components at a later stage in the project rather than installing everything together.

Sometimes it's the company that chooses to reduce the scope of the project to meet deadlines. Time to implement is becoming more critical to achieving business benefits. This sometimes involves reducing the scope of an early phase of a CRM initiative in order to achieve the business benefits and earn the right to continue with future phases. Systems integrators can be invaluable in such a situation.

Advice from the "Horse's" Mouth

Systems integrators offer the following suggestions to help a company get the most out of a CRM strategy.

First, have sound reasons and objectives for installing new systems. Fully understand the needs and know what systems can and cannot do to help the company meet its CRM objectives. Develop a blueprint that shows how systems support each other to provide consistent customer service. Don't expect tools such as IVR systems, SFA packages, or email management software to magically handle everything.

A company can have the best technology available, but if it isn't designed so that the customer's experience is the same across all channels the ROI will be limited. Eventually (sooner rather than later) it will be necessary to go back and retrofit, wasting 30% to 40% of the initial investment, not to mention the time spent on the original implementation.

Second, don't underestimate the CRM initiative's requirements. Avoid buying products whose capabilities will be quickly exceeded. Companies don't always save time or money when they purchase and install smaller or less expensive equipment and systems since they can require nearly as much integration effort and time as more robust products. If a company is looking at implementing technology to enable an advanced CRM initiative, the first step may be for an integrator to analyze and evaluate that customer's business objectives and growth patterns. For example, many integrators will conduct extensive studies of a company's customer base and discuss the business's short- and long-term goals.

It's just not cost-effective to buy low-end CRM technology that will be out of date in two years if a company's annual growth rate is 10% to 20%. A top-notch systems integrator will recommend CRM technology that will help a company meet its goals today *and* tomorrow.

Third, budget enough time to get everything up and running. Let's say a CRM initiative needs to be implemented by a certain date, and the anticipated implementation cycle is estimated to be six months. What happens when the contract talks and the approval of the funding end up taking the lion's share of the time, so that approval of the contract doesn't happen until (say) three months before the deadline, but the rollout must still occur on time? A squeezed implementation, items overlooked, and probable failure.

What to Expect?

The systems integrator should evaluate the CRM initiative's technology needs. From this the in-house staff (CRM team), the integrator and the vendor can generate a blueprint. This can take a few weeks to complete.

With the proper blueprint in hand, everyone knows what to expect. The typical installation and customization process is relatively straightforward and can usually be completed within a few weeks. Then the hard part begins — customer-facing and/or front-office systems must be integrated with back-office systems. For example, customer contact records generated by the sales department must flow from the CRM software into invoices, and the data from invoices needs to flow into account activity software. This is where systems integrators with the help of middleware earn their keep. Integrators map data elements from one system to their corresponding elements in another system, then use third party tools, which act as a translator, to allow data to traverse between the systems.

CAVEAT: Systems integrators that have heavy experience in the ERP field or other back-office solutions use a much different skillset than required for CRM. As such, they may approach CRM as a technology issue, and not a business strategy and people issue. CRM doesn't work when it's approached as a technology. As stated many times, CRM begins with strategy, and if using a systems integrator, the CRM team must find one that approaches CRM this way. Although the simple automation of business processes will improve operational performance, what many of this ilk fail to take into account is that automating already inefficient processes will only speed up dysfunction...a recipe for failure.

A systems integrator's pragmatic approach can deliver tangible results quickly. Many times, the integrator will be with the CRM implementation every step of the way. CRM integrators run the gamut — from the highly technical to visionary strategists. For example, some of the areas in which an independent systems integrator may be utilized during a CRM initiative might be:

* ✳ CRM vision and strategy.
* ✳ Database migration; data and database audits.
* ✳ Business requirements definition.

* Building a business case.
* Segmentation and data strategy.
* CRM systems infrastructure strategy.
* Systems and process preparation.
* Integration and implementation.
* Vendor selection.
* Preparation and management of service level agreements.
* Knowledge management.
* Project appraisal and performance reviews.

Systems Integrator Wrap-up

The real beneficiaries of the CRM explosion have been the systems integrators, from the large heavyweight to the small, niche integrator. Whatever the price of the CRM software, the integration costs will typically be five times that, and potentially considerably more.

Implementation of a corporate-wide CRM strategy is still leading-edge, and there currently exists a scarcity of technical and business experience within the business community to negotiate this transition. Many companies look to the CRM practice groups within the systems integrator community to help guide them through this process, and to provide the technical skills necessary to integrate disparate, best-of-breed "point solutions" with their legacy infrastructure.

THE SELECTION PROCESS

What should a company do when considering the possibility of hiring a tech partner? Formalizing an effective selection process is key to obtaining the best services for an exact need. As part of the process of sorting out the expectations when selecting a tech partner, write a summary of the CRM initiative including its objectives and needs. This summary can be a handy reference tool when making initial contact with perspective consultants.

Determine Need

To begin, assess the CRM initiative's specific requirements, review the objectives to ensure they are well-defined, then judge what the CRM team can handle and what needs to outsourced by determining:

* How much of the work can be done in-house?
* How much of the work is beyond the in-house staff's capacity and/or capabilities and must be outsourced?
* What is the budget?
* Are there desired cost savings? If so, what is the expected payback period? (i.e. ROI of hiring a tech partner vs. trying to do it in-house)
* What is the company's commitment to the project? (In other words, after

time is spent on the strategization and selection process, will the project move forward to completion?)
* What is the timetable for completing the CRM initiative?

Now compare the need to the types of services a tech partner offers. Including:
* A specialized expertise.
* Ability to objectively assess a specific situation.
* A temporary supplement to the in-house staff and knowledge base.
* Technical and economic analysis of alternatives.
* Development of recommendations.
* Design and programming support.
* Implementation and integration support.
* Assistance with hardware/software selection.
* Assistance with implementing operational changes.
* Completion of a one-time point solution.

Qualifications

Ask prospective tech partners for specific details regarding their expertise and work history. Request references and ask how the references are similar to the company's CRM initiative. Then check their references carefully by personally talking to a point person at the referenced businesses. Checking past work performance is one of the best ways to evaluate a prospective partner.

When checking references find out whether the prospective partner –
(a) has worked on projects similar in size and nature to the proposed project;
(b) met the stated work and project deadlines; and
(c) was responsive, available and trustworthy

On a more personal level ask –
(a) were there any problems and, if so, were they satisfactorily resolved;
(b) was it easy to work with the tech partner;
(c) was the tech partner knowledgeable; and
(d) what was the overall impression of the tech partner?

Also find out if the project came in within the projected budget. Ask if there were additional expenses billed. Find out if the referenced company felt that the final cost was in line with the original estimate.

The RFP

Once the prospective tech partners (whether consultant or systems integrator) are narrowed down to a short list of three or four, prepare a RFP. A RFP is a formal request to the prospective tech partner to describe everything the partner should accomplish and requesting the consultant to write a proposal outlining how it would go about meeting the specific needs of the CRM initiative and the costs thereof. The RFP can be as informal as an email or a telephone call, but it is best presented as a formal written document. The more concise the information it conveys, and the more specific the requests, the better the partner's ability to draft a relevant proposal.

There are basically two types of RFPs. The first is a defined, rigid proposal wherein is laid out, step-by-step, the CRM initiative's goals, objectives, and requirements. This type of RFP allows for an easy comparison of costs, approaches and other criteria submitted by the prospective partner during the selection process.

The second is a creative proposal, a three to four page document setting out what the job will entail. Such a proposal gives the tech partner less structure thereby allowing for a greater diversity in response, but this format makes the proposals more difficult to compare. However, the creative proposal allows the CRM team to observe how the experience and knowledge of the prospective partner can be applied to provide an unique approach to the CRM initiative.

Note: A RFP for a systems integrator should include a clear definition of the problems that exist with a current system, then outline the goals of a new system. It should contain a consensus of information from the technical staff to the employees that will be affected by the system. The majority of an RFP should be dedicated to business process, databases, systems and software issues since the company will want to make sure that a CRM initiative is compatible or can be made to work smoothly with the company's current technical architecture.

Review the proposals received from each partner group (consultant and integrator) and compare them to the established selection criteria to determine whether:

* The tech partner responded to the principal needs based on the RFP's outlined objectives.
* The services set out are specific to the RFP and are clearly defined.
* The timetable covers both the tech partner's time and company's employees' time and that it is reasonable.
* All fees and costs are clearly defined, the billing procedures are specific and the tech partner's fees seem reasonable.
* The tech partner has clearly defined the division of responsibility between the company's staff and the partner's staff.
* The tech partner sets forth a list of personnel assigned to the project, including resumes, experience and billing rates.

The Interview

After review of the returned RFPs and the CRM team has compiled a list from each tech partner group whose skills fulfill the necessary requirements, invite them in for an interview. Be sure to explain the CRM initiative's objectives and how the CRM team views the division of labor and responsibilities between the tech partner and the company's employees. Be specific. But also solicit the tech partner's opinions — an outside perspective is always valuable. Don't forget to set out the timeline (if it wasn't in the RFP) and to ask the tech partner to specify *exactly* who will be working on the CRM initiative.

Keep the IT department informed every step of the way. Make absolutely certain that the IT personnel are in agreement about the need for the consultant's and/or integrator's expertise, *before* signing the final paperwork.

Due Diligence

As with any potential partner, due diligence must be performed. Ask the tech provider for a list of at least five (preferably ten) customers with similar technical architectures and technology and/or consulting needs. Then question these references as to how well the tech partner met their needs and what would they do differently, if given the opportunity? This is the first step in conducting a careful due diligence process.

Next, look into the tech partner's financial viability, especially if the project is to run more than three or four months. Also determine:

* How long the tech provider has been in business.
* What kind of CRM projects the tech partner has handled in the past.
* What resources the tech partner has to complete the project, (i.e., staff, sub-contractors, etc.).
* If there are any foreseeable problems in meeting the envisioned time schedule.
* How the tech partner charges — per hour, per day, or per project — and if a deposit is required.
* If the tech partner historically meets milestones and deadlines.
* If the tech partner has a high turnover among its most talented people. Also determine the quality of the tech partner's staff.

Don't forget to investigate such issues as stability of management, information on physical plant, etc.

The due diligence process is the same whether it's for a tech partner or a vendor. So for a complete review of due diligence procedures, read the due diligence section of the next chapter.

Note: *In any communications with the prospective tech partner, endeavor to provide a written answer to any requested information — clear communication is important, and it helps to ensure that the company gets the results it wants.*

The Contract

Once a tech partner has been chosen, it's time to get the legal department involved since a written contract is a necessity. Some of the specific issues that should be addressed in the consulting contract are:

* Does the tech partner use subcontractors? If so, does the partner receive a commission for their services?

✳ Who provides the necessary insurance coverage? What type of coverage does the tech partner have?

✳ How will unforeseen costs be handled? Is the company's approval required before such costs are incurred?

✳ Where will the majority of the work be performed? If at the company's place of business, is there adequate space?

✳ If information accessible to the tech partner is confidential, insist that a nondisclosure clause be included.

✳ Include specific dates for the project milestones, reports that may need to be submitted, and the project completion date.

✳ What is the procedure for handling problems that might occur?

✳ How will revised work and costs be determined?

Also cover the extent of knowledge transfer that the tech partner will participate in and how any final source code will be provided to the company.

Work Plan

Most tech partners bill on a time and materials basis; therefore, to protect the interest of both the company and the consultant, it is necessary to set out a reasonable work plan with a not-to-exceed (NTE) cost. A work plan is like a blue print that maps out exactly what the tech partner has contracted to perform, with timelines and estimated costs at defined stages. Use the proposal received from the tech partner as a guide when drawing up the work plan. In some instances the work plan may need to vary due to the nature of the project, so a series of possible scenarios and costs should be built around best case, worst case and most likely scenarios.

✪ IN SUMMARY

Evaluate what partnerships are needed. Implementing a CRM strategy is a complex undertaking and, thus, will probably require multiple technology partners. Some within the business community will be able to implement their CRM initiative using only the CRM team, their IT staff and the vendor's team. However, most vendors can't address all of the implementation needs of a CRM initiative. The range of services and capabilities required is phenomenal — change management, data management, database architects, data cleansing specialists, web and Internet architects, communication specialists, call center technologists, and much, much more.

The CRM team needs to examine its own capabilities, the IT department's capabilities, the vendor team's capabilities and the level of dedication within these groups before determining what partners it needs to seek out. The partners need to understand the CRM strategy, the processes and systems, and the tradeoffs that must be made.

When negotiating a contract with a tech provider, chances are that the company will want a relationship stable enough to last three to five years. Therefore, everyone needs to understand that a long-term relationship won't be stable and healthy unless it's based on a win-win approach. If the CRM team tries to bully the tech provider into accepting terms that will actually result in it losing money, it will eventually backfire.

Chapter 16
The Vendor Selection Process

While many technical trends have short life spans, CRM has demonstrated that it has some legs. That doesn't mean, however, that all the vendors that have jumped on the CRM bandwagon understand CRM or that their software packages are, in fact, CRM solutions.

After the CRM team determines what the company wants and needs for its CRM initiative, the next step is to decide which software package will deliver. As the number of CRM vendors increase (with many claiming best-of-breed status from one angle or another), the CRM team must find a way to work through the crowd of clamoring vendors to reach its ultimate goal: the right combination of features and functionality, on the right technology platform. And do it without expending unnecessary time and effort. Any decisions made at this juncture advances the CRM team's previous hard work from theory to bottom line profits.

As word gets out that a company is looking for CRM technology, a feeding fenzy begins. A horde of CRM vendors will converge. As the CRM team listens to each new pitch, what once was clear can become very murky.

The team must be prepared — vendors want to make a sale; they will do whatever it takes to get a signature on a contract. Vendor sales teams are skilled at making anyone believe that if their product isn't chosen, the CRM initiative will fail. Exercise discretion and selectivity when scheduling meetings with vendors.

There are five items to keep in the forefront during the search for the right technology for the company's CRM initiative:

First, and foremost, CRM is in not a software application — it is a business strategy that can optimize profitability, revenue, and satisfaction at the individual customer level. Every single process and application within the company is a tool that can be used to serve a CRM goal or objective. The CRM products that the vendors are hawking must fit within this ecosystem — not vise versa.

Second, the software products that vendors sell will, in all likelihood, require customization before they will be of use to the company's CRM initiative.

Third, most vendors, if not all, sell their software products and their professional services, as a package.

Fourth, when it comes to the "flavor of the moment," vendors will go out on a limb to claim that they have the latest flavor, in this case, CRM solutions. But of the 500 or so vendors that currently claim to have a CRM product, only around 35% offer a true CRM solution.

Fifth, when looking for CRM vendors, realize that the typical vendor caters to one of three business categories: small companies (less than 100 employees) i.e. vendors such as Goldmine, Multiactive and SalesLogix; midsize Companies (100 to 500 employees) i.e. vendors like Applix, Pivotal and Remedy; and enterprises (more than 500 employees) i.e. Clarify, Oracle and Siebel. A CRM package designed for the small business class may be problematic for a larger company (even if only used as a point solution), since it probably won't scale well. Vendors that serve the smaller companies typically offer integrated CRM packages that lack the depth of those sold to the other markets. Also vendors that traditionally serve the enterprise class are venturing down market by scaling their product for the mid-sized market. At the same time, CRM vendors who've traditionally tailored their product for the mid-sized market are trying (with some success) to expand their sales territory by going after the enterprise market.

Plan the search process carefully. If not managed correctly, the vendor selection process can get out of hand — eating up time and resources — resulting in a less than optimal CRM solution. (The selection process can account for as much as one-fourth of the total costs of a CRM project.)

✪ UNDERSTAND THE CRM VENDOR COMMUNITY

To comprehend the potential universe of CRM vendors, the CRM team needs to understand that the majority of the current crop of CRM vendors came from a diverse category of technology providers.

First, let's discuss the *traditional CRM vendors* who typically offer the most comprehensive CRM packages (mainly through development of an end-to-end solution via acquisition and partnership strategies). Within this category there is a rush to consolidate, especially between various CRM vendors and call center solution providers. Mainly because a call center solution provider can (perhaps) give future leaders in the CRM vendor space an established customer base to build upon. It's noted that many of the "acquirors" are among the "who's who" of CRM vendors.

I would say that the **ERP vendors** run a close second to the traditional CRM vendors. This group has had its engineers burning the midnight oil so as to extend the front-office functionality of their back-office systems into full CRM suites. The ERP vendors who offer CRM products can link their CRM applications such as sales force automation to back-end support. Also, by leveraging their installed base of customers (who already use their database platforms), ERP vendors are able to easily (everything is relative, of course) tie, for example, analytical CRM solutions with their ERP back-ends.

Next are the *e-commerce software vendors* that market what they've classified as channel-specific applications. These products typically connect to the CRM related database, which, in turn, aggregates data from multiple customer channels and makes it available for front- and back-office users (call center agents, marketing professionals, sales people, etc.).

Finally, there are the *hosted solution providers*, as discussed in a previous chapter, that offer what are typically best-of-breed CRM solutions on a leased or pay-as-you-go basis.

Verticals

Some CRM vendors offer industry-specific suites that let companies deploy the software faster than generic packages since there isn't the need for a lot of customization work. These products can speed implementation and generate faster returns on investment.

By the time this book hit's the bookstores, Pivotal Corp. and PeopleSoft Inc. will have launched their first vertical CRM packages for retail banking and financial services, respectively. Oracle has already launched its CRM package for the aerospace, defense, and consumer packaged-goods industries and Siebel Systems Inc. offers industry-specific CRM packages for 18 separate vertical industries.

Note: Currently, the financial services industry is the largest adopter of vertical CRM technology. The second largest group to rush to the vertical CRM solution is the telecommunications industry; followed closely by retail, with media, healthcare and other industries at the rear (but this could quickly change).

A package that's vertically attuned to one industry will deliver greater benefit quicker, *assuming* best practices from that industry have been built into the vertical offering. Customers of generic CRM suites typically spend at least four months customizing 25% to 30% of the functionality; those who buy vertical applications spend only two months or so customizing 10% to 15% of the applications. With vertical integration already built into the product, companies should save on customization costs, which can account for up to 70% of a typical CRM rollout, according to industry analysts.

Vertical products aren't the panacea some people might think. Sometimes, even a vertical product can be too broad — the product might not be geared toward the company's specific business model, types of customers or sales strategy. Does the CRM team buy a vertical product that's built for the company's industry (say) retail, but isn't optimized for the company's customer base, e.g. businesses who buy office furniture that's custom built? This would force the company to re-engineer itself around often inappropriate and unnecessary features and processes. Or should the choice be a generic product that can be customized to "fit" the company's specific needs?

With such a wide range of diverse technologies offered by the CRM vendor community, most CRM teams will be lost if they aren't prepared for the selection process. The CRM team must look at both — vertical products and generic solutions. In the end, what really counts is how the product fits within the company's specific niche.

The CRM team should wary of inflated vendor claims about their industry-specific or vertical CRM solutions. Buy sensibly — get the vendors to be honest about their products' true capabilities. Everyone must realize that no vendor can promise everything for everyone — even with a vertical product.

Also perform due diligence. Designing and building a product that provides the necessary efficiency and effectiveness required by a specific vertical demands that the vendor has a deep understanding of that vertical market, redesigning a one-size-fits-all product doesn't fit the bill. Check out the vendor's background. Does it have expertise in the vertical market it is targeting?

Also the vertical product needs to be designed from the ground up for a specific vertical market. Look into the history of the product itself. Is it just a quick redesign of an existing product or a new product built from the ground up for a specific vertical market?

⊛ VENDOR SELECTION

Establish a vendor evaluation methodology. Business and organizational *needs* should drive the vender evaluation and selection process, rather than vendor *capabilities*. This requires ferreting out vendors that understand the company's size, class, industry, and that offer products tailored to that niche.

One of the great misconceptions that a CRM team faces is that CRM systems offer a defined set of functions as does, say, an order processing system. Nothing could be further from the truth. Any CRM initiative requires a carefully defined, and often quite limited set of functionalities, which are chosen specifically to support one or more customer-centric process. The only given is that each of these functionalities will feed, or be fed by, a single data storehouse containing data. Therefore, best practice requires that a well-defined set of functional specifications be established before beginning the vendor evaluation and selection process.

To help in the quest for the perfect CRM technical solution, the CRM team must embrace a few realities. First, the abilities of the software being evaluated;

and second, the importance of the company and the vendor sharing a common vision for the future. Next, consider vendors that have a strong post-implementation support team — these teams are vital for ongoing infrastructure, technical and process support. Finally, take it by stages.

Selection Criteria

Develop selection criteria (subjective and objective) around industry-driven features, vendor-related information, business vision and functions, pricing characteristics, a set of technical features, user friendliness/human factor requirements, and support metrics. Use the selection criteria to compare components, features, functions and technological capabilities that meet the CRM initiative's requirements. Try to identify both tangible and intangible benefits during this phase.

Define the Need

Complete a needs analysis by identifying the situation that dictates the need for a CRM tool, such as who will use the tool? What data fields and analysis do they need? What volume will it need to handle? What customization will be required?

At the same time draw up a requirements definition document (a scoping document that the greater mass of interested vendors can use as a starting point); be definite, but keep it concise and compact. It's surprising how many companies purchase CRM tools with but a scant definition of requirements in hand (with disastrous results).

Requirements Definition

Focus the requirements definition around a list of, preferably five, but no more than ten, major business issues, which identify the business process or rules that the CRM teams believes makes the company's CRM initiative unique. (A consultant may be helpful in building a consensus.) When completed, there should be a precise requirements definition that clearly outlines the "must have" technology for the CRM initiative. But the document should also provide:

* A concise description of the company, its goals and vision, its products and customer base.
* An overview of the CRM initiative.
* A wish list of desired (but not mandatory) functionalities, including those that might meet future needs.
* Short-term and long-term priorities.
* Other selection criteria (platform specific, database specific, and so forth).
* A projected timeline for the project.
* Resources the company intends to devote to the project.

Try to make the document as diagrammatic as possible, but avoid analysis paralysis. For instance, don't build an exhaustive list of mundane feature and function requirements.

A lengthy requirements definition can cripple the selection process — the critical business issues get buried in an avalanche of commonplace information. I've seen requirement definition documents (and RFPs) for CRM technology that could dwarf most books on the subject. Even a 25 to 50 page document can be overkill. Try to limit the requirements definition to fewer than ten pages.

Watch out for the poetic license some vendors take in their interpretation of certain line items; also avoid being lured into the infamous "with modification" trap. If not properly prepared with intelligence and forethought, requirements definitions (and RFPs), initially prepared to simplify the selection process, can instead collect a mountain of vendor propaganda and create confusion, rather than instilling a methodology that brings order to the selection process.

Map Goals and Milestones

The CRM team must establish a goal (or goals) and milestones (both short-term and long-term). Map out the milestones for each goal. Weigh any alternatives, such as the advantages and disadvantages of building an in-house infrastructure, versus going with a hosted solution versus doing a little of both.

Research

Research is the only way to identify the right vendors with the right technological solutions for the company's specific CRM project. While research can be time consuming; it must be done thoroughly. Begin by visiting vendor websites, reading industry literature, and attending vendor conferences and trade shows. Online communities are also very helpful and informative. Many companies in the market for CRM technology also use tech partners and research firms to evaluate potential vendors and their products.

In addition to looking at the better-known players, do a little homework and see if there are companies focusing specifically on the company's vertical market or niche. If so, that vendor's products may be a better fit for the company's CRM needs.

Narrow the Field

Don't invite a horde of vendors to meet with the CRM team or demonstrate their product. Cull the throng beating down the door to a list of no more than ten or twelve good vendor candidates that best meet the company's unique needs.

Talk to the vendors on that list to determine *exactly* how their technology package addresses the technical needs of the company's CRM initiative. Share with this group of vendor's the requirements definition document and the problems the team wants the CRM technology to address or solve. Then let the *vendors* tell the CRM team how the *vendor* would proceed to win the company's business. Intervene or guide where necessary. If the sales literature claims "integration," ask what the vendor means by "integration" — ten vendors will give ten different answers. It's the CRM team's job to decide which best fits the company's definition. If a vendor's response isn't satisfactory; cross it off the list.

Do not give the vendors a script to follow, let the vendors script their own solutions and then rate the presentation as part of the evaluation process. A key part of an evaluation process is how well a vendor demonstrates its ability to understand the company's CRM strategy and critical business issues and identify how their solution best addresses them, so don't give it away.

The idea is to find three or four vendors the CRM team feels comfortable with even though there haven't been any proposals with numbers yet. So, glean through the ten or so vendors and select a short list of companies that best meet the technical requirements of the CRM initiative. Invite them to bid on an RFP (which by now should be in at least a draft form). Don't give industry heavy-weights a free pass to the short list.

When a company plans to invest a large chunk of dough in CRM technology, it must do all it can to ensure it's making the right choice. Differentiating a total vendor technical solution on an equal basis when vendors present their products so differently is very difficult. Unless the CRM team makes a careful, fully informed decision about how and where it is delivering information to its customers, CRM tools won't do much in the way of helping it accomplish that goal. To illustrate: a CRM team may find that a vendor's product may address CRM in the call center brilliantly, but only pays lip service to wireless connectivity, or leaves out field service automation altogether.

The Short List

Never ink a deal until the CRM team has visited the vendor's premise and performed hands-on testing. But limit these items to only the vendors on the short list. These are much over-looked steps in the evaluation and selection process.

On-site visits offer the opportunity to share the company's clearly defined business needs, priorities, and key questions. A well-planned on-site visit should serve to shed new light on the vendor's products, services and solutions, while increasing the depth and breadth of the discussions between the vendor and the CRM team. Not a repetition of what's in the vendor's sales literature.

Send end users to a test site so they can test the software hands-on. Only hands-on experience can give a feel for ease, flexibility and functionality. Let the company's statistical analyst, campaign manager, sales professional and other significant users test-drive the stuff. After extensive testing, have these users rate the products and compare them to other vendors' CRM tools.

Scalability

Can the proposed system grow and scale as the business evolves? Many business-es will start their CRM efforts by focusing on only one part of their business, such as a point solution to address CRM in marketing, a call center or field sales. How easy will it be to integrate the vendor's technical solution into other department's systems as build-out of the full-scale CRM initiative progresses? While a single CRM application doesn't and shouldn't do everything by itself, the vendors should be required to show how the product can be integrated and scaled over time — with their products *and* with other vendors' product.

Ask how easy it will be for the company's staff to modify the CRM tools. As the company and its CRM initiative change, will the vendor's product keep apace and support any necessary shift in strategy?

Third Parties

Many vendors use third parties to integrate their product into the client company's systems and processes; but, without very specific terms in a statement of work, mixed results may rule. Third parties may not be as familiar with the company's business environment and processes as they should or need to be. It's the joint responsibility of the CRM software vendor and the company to carefully manage the success and quality of a third party's work. If it is left to the vendor to manage a third party integrator (for instance), risk to the project increases. Third party issues should be carefully considered before proceeding with negotiations. For instance, if the third party fails to do the work effectively, does the responsibility rest with the vendor? Carefully explore and negotiate all vendor related third party relationships.

Service and Support

A product's functionality and cost is what provides, in many instances, the inducement for making a specific vendor selection. But, it's just as important that the CRM team also consider the availability of quality service and support, for without them, the success of any product implementation is ultimately doomed.

Service and support need to be broken down into two areas. The first, *general support*, includes installation of the product, continuing support (regardless of location), and the quality of the help desk services. For instance, make sure the vendor provides access to user group meetings and/or forums, technical newsletters, technical updates and other information provided on (say) a "members only" website. This allows the CRM team to assess the vendor's effectiveness for ongoing support.

The second area is *professional services*. This can be the weakest link in many vendor organizations and therefore due diligence must be stringent. Determine if the vendor has a professional services group that can help provide the expertise needed to get the system up and running. If not, does it have well-established alliances with systems integrators or consulting firms who can provide that support? Evaluate the vendor's strengths in project management, systems integration, and business consulting skills. If the vendor relies on alliances, then check them out. What are their strengths and their weaknesses? Are they true business partnerships, etc.?

Training and Support

The vendor should be willing to help bring the staff and other users up-to-speed quickly. So, consider what type of help is available to train the staff on the new system. Does the vendor offer classroom training or computer-based training courses? Are the user guides easy to understand?

Evaluate the vendor's "hands-on" system administration and user training courses. Ask for a review of certified and context-based systems documentation. Observe the vendors' advanced technical training courses, including "train-the-trainers" classes, which should be combined with monthly training check-ups to ensure appropriate use of the software.

Then look at support going forward. When there is a problem with the system, whom does the company contact? What are the steps the vendor goes through to help solve on-going problems?

Check out support by performing trial runs via telephone, web self-service and on-site technical support. Don't forget to check out 24 x 7 help desk support and any other general support provided for the CRM software package.

These related services are as important as the software itself, and in some cases, even more so. When struggling to decide between two vendors, quality of training and support can be one of the key differentiators.

Focus

Another issue to consider is the makeup of the vendor's current customer base. A retail clothing business doesn't want or need a vendor whose customer base is dominated by, say, retail home furnishings businesses much less healthcare or telecommunications. Make sure there is a fit, i.e., the vendor's strategic direction matches the company's needs.

Pricing Proposal

Request pricing proposals. Every vendor has their own pricing structure, and it's hard to compare apples with oranges. Everyone knows this. That's one of the reasons the CRM team spends so much time and effort on requirement definitions and the RFP.

○ REQUEST FOR PROPOSAL

In spite of the critical service and support that vendor's supply, most businesses don't take the time to create a useful vendor RFP. Be the exception. Create an RFP that will attract well thought out and reasonable proposals and at the same time set the basis for a comfortable working relationship through what can be a complicated and expensive project. This is an area where a tech partner can be of help. While many of the company's employees may be comfortable working with RFPs, crafting an RFP for a turbine engine or a ocean liner is very different from drafting a RFP for a CRM initiative.

RFP Committee

Form a RFP committee. While there shouldn't be a huge task force (it would be hard to get meetings scheduled, much less get anything done in them), it's advisable for the CRM team to invite serious contributions from a wide variety of sources, including:

* Marketing.
* Sales.
* Field services.
* Investor relations/corporate communications.
* Customer service/call centers.
* Web operations.
* Order processing/fulfillment.
* Finance and legal.
* Outside service bureaus and service providers including marketing, advertising and PR firms.

Each of these areas have data that should be considered, and they'll all expect, in some way, to touch the data that comes out of the new integrated system. Their input is needed when drafting an RFP.

Keep the IT department happy. Make it clear that their concerns, along with everyone else's, will be addressed.

Drafting an RFP will take longer than any estimate. In fact, it could easily take three months, even with the help of an experienced consultant. Meetings are hard to schedule. Then there are holidays, weekends, vacations, and "things" just happen. (Note that smaller projects, limited in scope such as a single point solution, may take less time.)

Criteria

Don't, I repeat DON'T, provide an RFP with check-off columns where potential vendors can say whether they comply or not; this type of RFP is useless. A check off box is like asking someone "do you speak another language?" If the answer is yes, you still don't know what language or their competence.

Instead, ask that vendors provide screenshots or a detailed explanation to support every single answer given — it's the best way to ensure that the vendor really can provide the technology required. This type of proof should be a non-negotiable issue. If a vendor can't or doesn't provide explicit information, don't schedule a pitch meeting — mark that vendor off the list.

CAVEAT: *Many CRM vendors claim their products can do everything, or close to that — they can't and they don't. Be particularly alert to vendors that state the company's processes and workflow must follow the way their systems are designed. A couple of major vendors are notorious for this practice.*

For a RFP to be of benefit it must clearly speak to the CRM initiative's needs. It's easy to get swept up by information on hot new technologies and vendors; but the starting point for a good RFP is the CRM initiative's unique needs. The RFP committee should detail the following basics before beginning to examine solutions:

* What should the CRM solution do? Or what problems should it solve?
* What is the expectation, i.e. how will a product enable the CRM initiative?
* How will the CRM component(s) interface with the various users — internal and external?
* Determine the minimal "ease of use" features.
* What integration issues need to be considered? For instance, should it have the ability to integrate with specific hardware and software?
* How scalable should the product be?
* Does it adequately address the company's security concerns?

An Example RFP

When drafting a RFP break it out into sections. The following is an example of the typical RFP layout. It IS NOT intended to be an example of an all inclusive RFP. Use the following ONLY as a guide to what a RFP should consist of.

It's virtually impossible to include everything in an RFP, so don't try — the RFP will become a heavy tome rather than an informative document. Draft an RFP that sets out the way the two entities and their employees will work together. Remember the RFP rule: Everyone almost always promises they can deliver everything needed for the project in a neat little pile at the very beginning of the project. It never happens that way, so plan for the inevitable — the RFP should state what will happen when *either party* is late with a deliverable.

1. *Vendor Information*: This section requests a brief description of the vendor. It should cover:

* The basics: Vendor's name, primary address, main telephone number, URL and email address.
* Corporate detail, including financial information. An example: Is the vendor publicly owned? If not, give the name(s) and addresses of principal owners, as well as percentages owned. How are they financed? How long has the vendor been in business, and what is the history of the business?
* Vendor experience. For instance: Does the vendor have experience with the company's particular industry? If so, provide details. How many software systems has the vendor installed? Of that number, what percentage of customers remain active and what percentage dovetail with the company's class, industry and/or niche?
* Whether the specific product under consideration has "seen battle." Ask about how many installations of the product are underway or completed? It's important that the specific product under consideration "has been there,

done that and everyone's lived to tell the tale."

* The vendor's technological direction. Is it, for example, a web strategy, CRM modular approach, or ...?
* The management team. Who are they and what are their backgrounds?
* Reference. Ask the vendor to give a minimum of five and, if possible, ten references that match the company's class, niche and industry and that have implemented a system similar to the company's proposed configuration. Require that the information provided include each referenced company's name, a contact person and telephone number. Also, the identity of the system, i.e. its model/release and configuration.
* Product distribution method.
* A chronology of product development and releases. This should cover at least the past three years.
* Release information. Such as, how many releases have been issued since the product's genesis? How often are new releases issued? What's the approximate cost of new releases?
* Contact information. Who will be the initial and ongoing contact(s)?
* Asset and draft delivery methods. This should include project stages, milestones, quality control and testing.
* Other products provided by the vendor other than those specifically requested in this RFP. Ask for a brief description of each.
* Additional services. For example, does the vendor assist in initial or ongoing analyses to improve performance beyond the normal system training? If so, describe the services available and general pricing methodology.
* Team statistics. Ask the vendor to provide a complete description of the team (including qualifications) that will be involved in the project and assigned to the company's CRM initiative, including resumes.
* Vendor's on-site requirements. Ask the vendor what facilities and equipment it will need the company to provide while it's people are on the company's premises.
* Implementation schedule. Ask the vendor to provide a proposed implementation schedule with details as a starting point for drafting a project timeline.
* Costs and payment. The vendor should provide a detailed description of the total cost the vendor will charge, including post-implementation charges. Ask for payment details.
* Terms and conditions.

2. *General Information*: This section should include general information about the company and a concise description of the purpose of the proposal, but also:

* Details on what is expected in the software demonstration.
* Who is responsible for the costs of responding to the RFP.
* Number of copies required of the response document.

* Instructions on completing the RFP.
* The proposal evaluation period.
* Criteria for selection.
* Use of subcontractors.
* The company's typical payment method.
* The negotiation process.
* The company's legal requirements concerning non-disclosure and confidentiality issues.
* Marketing and publicity restrictions.
* Contract stipulations.
* Execution requirements.
* Contract cancellation provisions.

3. *Background Information*: This section should provide a concise, but detailed background of the company's need for the CRM technology. A brief outline of the CRM initiative is sufficient.

4. *Requirements*: This section should provide the specific technology required to enable the stated CRM initiative. This is where the RFP committee should describe, in detail, the technology requirements. Also, within this section there should be a detailed outline of the company's current technology architecture.

5. *Software Capabilities and System Requirements*: This section is where the RFP committee asks the vendor to describe the basic and optional components of the vendor's product(s). The vendor should generally describe the role of each component and its interaction with other components. Questions in this section might include:

* Is the vendor's system compatible with the company's technical architecture, as outlined?
* Is the data stored and made available in real-time? What is the vendor's definition of real-time?
* Describe how raw data from the vendor's technology is transferred to other systems. Is additional hardware or software required?
* Does the vendor's software have the ability to interface with other software packages? Give details.
* What type of database does the software require?
* Is the system restricted to a particular hardware platform, operating system, or processor? Under which operating systems does the software run? Are there additional hardware costs for non-standard operating systems?
* Does the software conform to the most recent common user interface standards? Is it mouse-driven?
* Is the system modular in design? Can additional modules be added later?
* Is the system password protected? What are the levels of system security available?

✳ Is it possible to access the software from outside the corporate walls? If so, what are the capabilities provided?

This is where the RFP asks for specifics, such as:

✳ Describe the system's reporting capabilities. What customizing features are included in the report package? How many standard reports are provided?
✳ Is it possible to archive reports?
✳ Is it possible to name reports?
✳ Does the system have any automatic reporting capabilities?
✳ In what time increments can reports be generated?
✳ Are bar, line, three-dimensional, and color graphs available?
✳ Provide samples of all available reports.

Other specifics that might be covered in this section would be, for instance, forecasting.

✳ Describe the forecasting model.
✳ Describe how the forecasting mechanism is able to trend normalized averages against historical actuals.
✳ Describe how current data is "weighted" when averaged with the historical trends. Can the weighting be user-defined? What assumptions are made in cases where there is limited or no historical data available?
✳ What data items are used for forecasting? Can they be user-defined?
✳ How does the vendor ensure that incorrect or inconsistent historical data do not corrupt existing trended files?
✳ Describe how a forecast can be changed prior to the specific day.
✳ Can the user modify a forecast?
✳ How long is historical information stored? How far back in history will the forecasting software reference data?

Note: *The specifics covered in this section are dependent upon the needs of the specific CRM initiative.*

6. *Hardware*: Since not all software works with all hardware, the RFP committee must ask the vendor to:

✳ Describe any hardware required to run the proposed system.
✳ Does the vendor provide this hardware?
✳ If PC-based, is the PC dedicated to running the management software?
✳ How many workstations does the system support?
✳ Does the product operate in a LAN environment in such a way that all workstations on an existing LAN can access the software?
✳ How many users can access the system concurrently?

7. *Training*: This section is of vital importance, some of the information that should be asked the vendor during the RFP process:

✳ Describe the training provided. Is separate training provided for the techni-

cal staff, management, end user and trainer? Are there any prerequisites?
* How many users are included in the standard training package?
* Is additional training available?
* Does the vendor support web-based training and conferencing?
* How many training personnel are available for an installation of the type proposed?

8. *Maintenance*: This section should be specific. For instance, ask the vendor to describe in detail:
* The maintenance agreement.
* The warranty provided by the vendor. Does it include software upgrades? What is the warranty period and what is the bug-fix policy during this period?
* The support and maintenance program. What features are included in the maintenance fee? Can users get telephone and online support from Help Desk personnel? If so, are both 24x7?
* The vendor's upgrade policy? How often is software upgraded? Are new releases compatible with previous releases? What is the process of interaction required if a specific hardware/software is changed or upgraded (an ACD for example) after installation of the proposed system?
* Background information regarding resources available to support the proposed system. Determine if the vendor or third party provide these resources?
* How the vendor manages ongoing communication with its customers regarding new releases, additional training, user meetings, etc.

9. *Cost*: This section can cover quite a bit of territory. How much should go in a RFP and how much should be negotiated later should be determined on a case by case basis. However, I provide the following as a sample.
* Application software costs.
* Packaged applications costs.
* Custom applications costs.
* Modification charges.
* Development cost for system requirements that are not currently available.
* Warranty, maintenance, and support fees.
* Itemized hardware costs.
* Installation charges.
* Maintenance prices.
* Is software leasing an option?

10. *Miscellaneous*: There will always be something that doesn't fit within any specific section. For instance the company may want to request:
* A list of available publications or documentation.
* Other literature that's appropriate to the proposal.
* A sample contract.
* How the vendor implements its software?

* How the vendor provides the source code? Upon completion, via escrow or ...?
* A sample Service Level Agreement.
* How important is this piece of business to the vendor?

11. *Technical Selection Considerations*: Some items just won't easily fit in "Section 5: Software Capabilities and System Requirements." Use this section for those items. As illustration:

* Does the vendor use object-based architecture? (Or perhaps, to stress the fact that the client company uses object-based architecture.) This is a very important criteria to iron out. The object-oriented design approach of using COM/DCOM and CORBA technologies like ActiveX and JAVA controls facilitates integration with third party software and enables additional data and functionality. However, both parties must support such architecture.
* Many CRM initiatives will require computer telephony integration (CTI). Although CRM software vendors are increasingly offering interfaces using this trend, if this is a requirement don't assume, ask.
* Handheld devices are increasingly a standard computing device within the corporate environment. These increasingly sophisticated handheld devices are able to synchronize and store CRM information and communicate wirelessly. Does the vendor's product support such devices? Can the product easily support wireless communication if this function is requirement in the future? If so, completely, partially or only with the introduction of third party products?

12. *Implementation Schedule*: This section should consist of the CRM team's tentative schedule. List dates for the beginning and ending of the project and tentative deadlines for completion of the major tasks of the project.

The RFP should also set out a schedule for the proposal and development process. There also will be other milestones that will be set out once the project gets underway.

There's plenty more that can go into a RFP. Just keep in mind that what the RFP should accomplish is the selection of a vendor everyone can trust. For example: During the development process everyone will make mistakes. The company will be late with deliveries and so will the vendor. The best bet is to select a vendor who will forgive the companies peccadilloes, as the company will forgive the vendor's. The RFP should focus on the details of how the parties will relate to each other during the implementation process.

Provide the prospective vendors with a contact person whom they can talk with when they have a question about the RFP. That person should be "on-call" to answer vendor's questions during the time the vendors are working on their proposals.

⊙ PRE- AND POST-RFP TASKS

Before releasing the RFP, several tasks should be completed. First, establish a grading system for properly evaluating the proposal responses. Try to assign numeric values to as much as possible. (A matrix format, for example, could be created setting out criteria and standards to be used in compiling and rating vendor responses.) Judge the completeness of the information provided. There should be no ambiguities. Don't be afraid to thoroughly discuss and debate the shortcomings and strengths of each vendor's offerings and the alternatives available. Don't make a decision without a consensus. (When subjective grading is needed, make sure this is done in a team meeting and that notes are taken.)

The objective is to ensure the vendor proposals provide a means to achieve an apples-to-apples comparison. Prepare and send the RFP to all of the vendors on the short list. Products and prices can be compared more effectively when all vendors provide information using structured guidelines.

After the RFP is sent out, allow adequate time (at least one month) for a response.

Legalese

Talk to the legal department to determine if the RFP should be identified as a private, copyrighted document that may not be shown to others. Discuss confidentiality and non-disclosure agreements. The RFP lists detailed items about the company and its IT infrastructure. Take every precaution to keep that information confidential. For instance, many vendors use freelancers and part-time help, which isn't necessarily bad, but a confidentiality agreement can help to ensure the RFP doesn't get circulated all over town.

Include a copy of the company's standard software license agreement (if available) along with the RFP, and require that each vendor send its standard software licensing agreement as part of the RFP submission. Every company should ask that the vendor submit a specific number of copies of the completed RFP so there are enough copies for the company's evaluation process.

CAVEAT: DON'T ever release the budget for the CRM initiative or point solution.

○ AND THE WINNER IS...

The CRM team can now take the final steps — selection and negotiation. It's pretty straightforward. To whittle down the vendor selection list use RFP validation, comparisons, vendor presentations and references. Then schedule a no-holds barred in-house meeting and review everything to determine which vendor better fits the needs of the CRM initiative.

Choose the two finalists that most closely meet the CRM initiative's needs. Have both conduct a live demonstration and provide real-world product demonstrations. If possible, provide samples of real data (record data, retention information, user information) for the demonstration. The entire CRM team should attend the demonstrations and encourage the IT members to barrage the vendor with specific, detailed questions regarding the product and how it will work with the company's technical infrastructure.

Note: *Some vendors request a list of pre-determined questions prior to a demonstration, which is just fine; but make sure the vendor understands that ANY question may be asked.*

When the demonstrations are complete, have the CRM team meet as soon as possible to select the winning vendor product. It's better to make the decision while the information is fresh on everyone's mind. (An audience that views information in a "show and tell" format — the typical demonstration mode — will retain up to 85% of the imparted information for three hours but there is only a 65% retention rate after two days.)

Costs

Don't put too much emphasis on the initial cost of the product during the decision-making process. Substantial discounts abound —- license deals (whether a per-seat, per server, or per site basis), delayed maintenance charges, gratis training, and so forth; but don't be swayed by discounts. Remember that the majority of the cost is hidden — such as product training, customization and integration. Therefore, cost calculations should always include not only initial license costs for the product and any knowledge tools and middleware utilized, but also costs of data preparation, installation and maintenance, help-desk gateway, ongoing education and training, and professional services such as customization and integration.

The prices for CRM software are all over the place. And this is one arena where a company doesn't necessarily get what it pays for — companies achieve great successes and spectacular failures implementing both inexpensive and very costly solutions. The key is to find the system that is right for the specific situation. Then determine what the return on investment will be after successful implementation. If the ROI is significantly higher than the cost, buy it — it's not the price of the CRM tool, but rather the cost of ultimate ownership that's at issue.

Vendor Vision

A vendor's vision is just as important as the vendor's ability to produce a viable product and execute its successful implementation. For most businesses, this can be used as the key differentiator when making the final vendor selection.

What is the vendor's own stated and realized development plans for its product? For example, ask how the vendor plans to incorporate and utilize new technologies and how it plans to evolve its current product by adding to or enhancing the current functionality. Let's not forget about service and support. How does the vendor plan to grow and change its general and professional services support?

Due Diligence

Performing proper due diligence has proven to be the weakest area in most vendor selection processes. As with any potential partner, it is imperative that due diligence be performed.

References

A final decision criterion is reference checking. Check references — ALWAYS. Ask vendors for a complete list of their active customers in the company's niche (size, industry, scope of CRM project and business rules) so the CRM team can question these references as to how well the vendor and its products meet the needs of a live, industry-specific customer. This is the first step in conducting a careful due diligence process. The vendor should be able to provide at least five and preferably ten references that fit within the company's specific niche.

Reference checking does more than give insight into a vendor's capabilities; it provides priceless, real-world advice. The intelligence obtained from talking to other companies that have completed implementation of the vendor's product(s) is invaluable. It's impossible for a CRM vendor to develop the perfect product for everyone; and it's impossible to avoid surprises during the implementation period. Heed the advice obtained from the vendor's references — they've already "been there, done that." Ask the references how easy was it for them to modify the software? How well did the vendor support them? What glitches did they run up against during the implementation and/or rollout process? How did they and the vendor handle these problems?

The most important question you can ask a reference is "What would you do differently?" Then listen and learn. Remember to investigate how the vendor treats its customer after the final check has cleared. Also talk to businesses that are in the midst of implementation. The information gleaned from these companies can be very helpful to the company's CRM efforts.

Don't just contact two or three references but rather call as many as the vendor will provide. If the vendor can't provide at least five client companies in the same market space as your company — find another vendor.

Note: *Reference interviews are like free consulting — the information acquired will save time, resources and money, which will justify the time and effort the CRM team expended to contact them.*

Vendor Viability

Given the fierce competition that exists within the CRM vendor community, the author predicts that within a few years at least half of the current vendors will no longer exist. Has this been taken into consideration when making the vendor selection? If not, do so. What is the vendor's financial viability? This question can be answered through analyzing a vendor's revenues, growth, margins, sales and marketing investment, quick ratio, etc.

The next step is to measure the quality of personnel within the vendor's sales and development departments. Ask questions, such as do these departments historically:

* Meet industry milestones?
* Meet time deadlines?
* Deliver what they promise?
* Support R&D capabilities comparable with other vendors?
* Have a high turnover among their most talented people?

Also, carefully investigate issues as stability of management, and status of their physical plant (lease, ownership, condition of building, equipment, etc.).

◯ NEGOTIATION

Send acceptance and rejection letters by the date stated in the RFP. The CRM team may want to let the second place finalist know that if finalist one doesn't meet contract, it will move to finalist two.

The final step is negotiation. What the vendor set out in the RFP isn't written in stone. Start by reducing the total cost by 25% and when discussing ongoing maintenance and support, never pay an annual fee of more that 15% of the total software cost.

Paperwork

This is another stage where the team sharpens its negotiating strategy. How? Decide what's critical to the CRM initiative and put it all down in a contract (use the legal team for this part). Call in the selected vendor and the race is on.

Always tie payment to performance — from pilot projects to full rollout. If the CRM team can, negotiate a contract where payment is due when the system is fully functional — and the vendor is responsible for getting it there.

Note: *This is the best time to get concessions. It's the only time the company will be in a position to make demands, so take advantage. Once the contract is signed, the company is no longer in the driver's seat.*

After the vendor agrees to work within the negotiated terms and conditions, it's time for the final documentation, presentation to management and getting signatures on the dotted line.

Following the procedures outlined in this section enables a CRM team to utilize the best evaluation criteria, gather the necessary and objective data, and guarantee that the overall evaluation process proceeds in a structured format to selection of the optimal vendor. The CRM team will also end up with a great paper trail, which can be useful if it ever needs to explain how it arrived at its selection.

Let legal counsel take care of all paperwork going forward. Once everything is signed, it's implementation time.

⊙ IN SUMMARY

Today's customers demand superior service. This, in turn, has caused the business community to push the CRM vendor community for better tools. No longer can vendors simply promise the benefits of CRM. They must make a persuasive case for how their product will actually deliver those benefits.

Many times, the gut reaction is to go with a big name vendor, mainly because they are perceived as less risky. But, better practice is to analyze the technical needs of the stated CRM strategy and then make the technology decision. It may be that a melding of several different vendor products are better than going with one vendor's solution.

At this time, no one vendor can cover all of the areas of the typical full-scale CRM initiative. Also never forget, technology doesn't work in a vacuum, it supports what people do (i.e. workflow) and once the workflow has been figured out, it's possible to know what the new CRM technology will be able to do.

After the contract negotiations are completed and the signatures are on the appropriate documents, the next step is to develop a system migration plan, as discussed in the next chapter.

Chapter 17
Implementation and Deployment

A successful CRM implementation and deployment is one that is designed to fit the corporate culture and then is systematically implemented throughout the corporate environment. Tackling a corporate-wide CRM strategy is complex, expensive and even a bit risky. Even when implementing a point solution or a pilot project, there is the desire (and even need) for a speedy implementation process not only within the executive boardroom, but also within the CRM team itself. Everyone wants tangible returns from the project. After all, tangible returns are the best way to obtain funding for future CRM projects.

While the number of successful CRM initiatives is growing, some CRM initiatives still fail. Most of the blame can be placed directly on deficient business process re-engineering. Yet, other culprits can share the blame: a misunderstanding of the objectives of the CRM initiative; incompatible IT systems; lack of expertise and resources; poor integration; and confusion over technology, in general. Some of the burden for failure can also be born by lack of prerequisite technology in some corporate IT architectures — back-end database consolidation, VoIP networks, and wireless user applications, to name a few.

Note: A 2001 Alexander Group (a managing consulting firm) survey found that a business can regroup after a troubled CRM project. Fifty percent of companies whose initial CRM project failed have taken action. The results: 25% of this group are experiencing significant improvement. More notable is that a whopping 80% of these initially troubled projects have successfully relaunched after the CRM team's focus was redirected toward emphasis on process redesign, change management, and performance management.

⊙ IMPLEMENTATION PHILOSOPHY

Thank God, most CRM implementation and deployment hasn't followed the "swallow the whole cow" philosophy of the ERP generation. Lessons have been learned and CRM implementation and deployment is following a more sane approach — chopping the cow up and eating it, one steak at a time. (A few CRM experts actually refer to the typical CRM implementation process as "chunking." I like my analogy better.) This refers to structuring the CRM strategy into incremental initiatives, beginning with a pilot project or point solution that has a potential for success (a campaign management system, for instance); i.e. one steak at a time. Then move forward with other point solutions, applying a systematic implementation of the corporate-wide CRM strategy.

This approach makes sense for a number of reasons:

* The people factor, i.e. the employees must learn the new process after it has been re-engineered (some will eventually learn additional processes as the CRM initiative evolves).

* Implementation by stages and then re-evaluating the process before advancing to the next stage of a CRM strategy can help to reduce the overall risk of a corporate-wide CRM initiative and improves the success ratio.

* Technology changes, advances, and ways of implementation and integration will continue to change over time.

The Typical CRM System

The typical CRM system (not strategy) can be defined loosely as an integration of front-office customer contact systems and back-office operational support systems.

Yet, a CRM system also can be defined as a system that automates some portion of the front-office customer communication channel to the back-office operational applications with the ultimate goal of a holistic integration of all corporate IT systems and applications. This includes all systems that handle interactions with the company's customers (billing, customer service, field service, marketing, sales, etc.) as well as operational support systems (work order management, customer information, and network management information systems).

⊙ THE IMPLEMENTATION AND DEPLOYMENT PLAN (I&D PLAN)

The corporate-wide scope of CRM involves far more than simply installing a front-office application. Information must be collected from diverse sources, requiring linkage between disparate IT systems. Therefore, an I&D plan is critical. It's, in essence, a blueprint of the overall implementation: project phases, timelines, milestones, identification of resources, procedural changes, and assignment of due dates to tasks.

How does a company begin to develop what can be a formidable I&D plan? First, form an implementation team. Second, draw up and implement a systems migration plan. Third, during implementation and deployment break large projects down into manageable pieces. Fourth, provide for proper training courses. Fifth, stay on top of everything — don't let anything fall between the cracks.

The Implementation Team

Who should implement CRM systems and processes? An implementation team that is composed of three different disciplines: *strategy, business process redesign*, and *technology implementation*.

Begin with *strategy*. This requires someone who understands front-office tasks, the staff's roles and their needs. This can be either someone in-house or a consultant.

Next is *business process redesign*. This is required to optimize the existing systems and to accommodate the proposed CRM initiative. Data cleansing in preparation for data consolidation also belongs within this discipline. Again, most of this can be accomplished via in-house staff; but perhaps an outside partner would be a better choice for some of the tasks.

The final discipline is *technology implementation*. This requires teamwork between in-house personnel, the vendor's staff, and other outside partners who are onboard. This is where selection of the most appropriate technical system to match the CRM initiative's needs takes place. Here, I would suggest the CRM team and in-house IT representatives partner with a consultant and/or systems integrator, especially if the company already has either partner onboard.

The CRM team and implementation team can use the project plan and requirements definition and vendor's proposal as guides when drafting and finalizing the I&D plan. One of the main objectives of the I&D plan is to schedule and track the installation, configuration, and customization of all CRM technology (hardware and software) whether a point solution from one vendor, a product suite or CRM products from multiple vendors. The CRM team can provide task estimates and input regarding delivery dates to help drive the milestones and implementation of a schedule. But there should be input from many sources during this phase. For example, each department touched by the CRM initiative must help to determine due dates for integration of any data sources the department currently uses.

It's recommended that a single contact from each department affected by the project be available during this stage to provide content, answer questions, and give feedback. (Note this person can be, but normally isn't a member of the implementation team.) There should also be a primary project contact person appointed. That person should have decision-making power and can be the project manager or program manager (but it can also be a senior member of the CRM team).

The Systems Migration Plan

Let's just get this out of the way now: systems integration is hard, risky, and no matter what anyone says — it's *never, ever seamless* straight out of the box. There are a host of *non-technical issues* that arise when installing CRM technology. These issues can arise before, during and after the integration process.

Although the systems migration plan is a subset of the I&D plan, it actually provides the second stage of a successful I&D plan. The purpose of developing a systems migration plan is to enhance applicable processes and to identify implementation constraints. An ideal plan lists the hardware and software requirements, upgrades and restrictions; data requirements; and high-level modification requirements. It also provides a project priority list; sets out training needs; and maps out a conference room test pilot.

The systems migration plan is the basis for the I&D plan. For instance, a system migration plan would determine how to integrate the CRM system with call center ACDs and/or IVRs or a web front-end, as well as ascertaining how to move the data to and from data storehouses.

Depending on the complexity of the company's IT architecture, a systems architect may be needed to help with the CRM systems migration plan. (For example, while the actual process of moving the data will be the responsibility of the database administrator, the systems architect can determine the best method.)

This is when to determine whether systems integrators (or additional integrators) may be needed to hook the CRM system to other systems. CRM systems must be integrated with other applications and systems — from groupware to databases to ERP systems.

The Three Basics of CRM Integration and Implementation

There are three basic technical areas that CRM integration and implementation teams must consider:

Communications. This refers to the common framework that enables the necessary applications to communicate — the "ties that bind." This framework holds the CRM system together.

Information. This refers to the data that comes from many different sources. It's the means whereby customer intelligence is created. But building a CRM system that can access multiple data sources isn't easy. Basically there are two approaches that can be taken. One is to introduce a layer of software (middleware) that acts as an interpreter; receiving and interpreting data from the data source(s) and then interpreting it for each requesting application. The other approach allows the CRM applications themselves to understand what the data means, thus an interpreter isn't needed, e.g. EAI and XML.

Business and Work Processes. This refers to a complex set of interacting processes that must work in tandem in a seamless manner. Even relatively simple actions, such as processing a customer's order involves complex business and work processes. The implementation team must understand what processes are required to fulfill a customer's order. If, for example, a company doesn't have a way of triggering a business process, which, in turn, triggers a work process to let it's customers know about (e.g.) product shipment delays due to massive flooding, then the company is operating with a crippled CRM system.

Conference Room Pilot

One task on the systems migration plan is a "conference room test pilot." Since a conference room pilot is designed to suit a CRM initiative's unique circumstances, this book only provides a guide. The purpose of a conference room pilot is to allow the implementation team to see that the CRM technology can support all that the CRM initiative requires.

Develop a scenario that represents what the CRM system should be able to handle on a daily basis. The test pilot should be able to not only test draft procedures and policies, but also to identify:

* Any new procedures that might be necessary.
* Differences in business process and systems functionality within the current technical architecture and the future technical environment.
* Training needs.
* Implementation method and issues that need to be resolved.
* Data needs (cleansing, data definitions, conversion rules, etc.).

The test should identify most, if not all, of the major issues that require resolution. The test and resulting gap analysis serves to document and match current business process to changes that must be made to accommodate missing functionality in the new system design and the agreed-upon modifications.

Process Re-engineering

Many within the business community have been through some kind of process re-engineering before. But that was small potatoes compared with what must be done for a successful CRM initiative. Just about every process that touches CRM must be re-engineered for the customer-centric environment. Even the customer-facing processes in place (sales, billing customer service) may need updating or replacement. Thus, the implementation team must look at the way the current processes work (or don't work) and then re-design or replace any faulty process. Even when best-of-breed CRM technology is brought onboard, if it is integrated with a flawed process, the end result is a flawed CRM initiative.

CAVEAT: *DON'T try to correct a process deficiency by purchasing a CRM application that the vendor **claims** to have been pre-built to provide the necessary process(es). The application must be customized to the company's specific requirements.*

Complicating matters is that few in-house personnel, even IT staff, can read a process map, much less understand it. So before re-engineering any existing

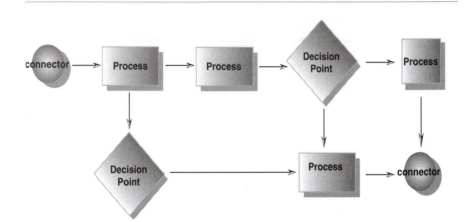

Figure 30. A basic process map graphically depicts the sequential steps involved in converting a specific input into a required output. This type of mapping helps to direct and focus analysis and change initiatives by uncovering the nuances in a process. It points out the "moments of truth" in both decision-making and customer service.

processes, create a diagrammatic document or documents clearly depicting, for example, the customer information flow. Then support it with quantitative and qualitative data.

The process map should enable everyone to understand how any process currently works and what needs to be changed prior to implementation of any new CRM technology. As an example, some the items a process map can help with is:

* Documenting key process steps.
* An understanding of the process needs from the perspective of the customer-facing staff.
* Identification of any gaps in process functionality.
* Determining actions that should be taken to enable the process to work effectively within a CRM environment.

During this period, the implementation team should meet with key customer-facing staff to obtain feedback on the various processes. Compare the results with the process map. This will help the implementation team to categorize and prioritize the necessary re-engineering work. A consultant with process re-engineering expertise might be invaluable during this phase.

Note: *To help non-technical staff members to better understand what needs to be done, create a process map using clip art or rough hand drawings to represent computers, printers, fax machines, email servers, the Internet and web channels, brick-and-mortar locations, incoming information, outgoing information, etc.*

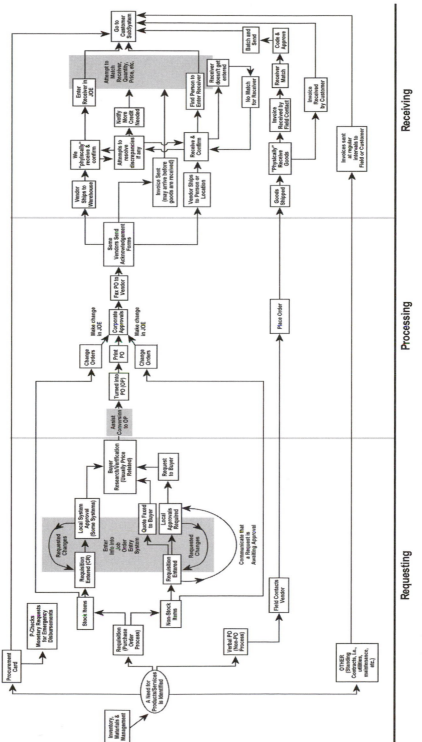

Figure 31. An example of a purchasing process map. Courtesy of the Academy for Corporate Excellence.

Ideal Methodology

Before installing CRM enabling technology (or even before buying it for that matter) the CRM team must evaluate the company's business process. If there are any flaws, the process must be re-worked and corrected.

CAVEAT: *Starting with a legacy process and **attempting** to re-engineer it can result in a new process that doesn't accomplish much — **unless** the re-engineering work is intelligently thought out and implemented. The best methodology in dealing with legacy (especially mainframe) process may be to use a third party (consultant and/or integrator) who can evaluate the company's processes and needs, and give a fair assessment — most companies can't and don't look at themselves objectively.*

Realistic Methodology

While it's always better to have the process re-engineering work completed before installing CRM, for many businesses, that's not realistic. The longer a company must spend on reorganizing and testing its process the longer it must wait before it can install the new CRM technology, let alone realize the benefits.

Many within the business community, therefore, have decided to not wait until the company's entire business and work process has been re-engineered (this can take months, if not longer) before installing CRM enabling technology. They deploy CRM and its technology on a department-by-department basis (i.e. point solutions).

It's *possible* for a company to concurrently implement CRM software *and* process changes, process-by-process, *but only if* a consensus is reached on the strategy and everyone is prepared to make the required organizational changes in each function. Departments involved in the CRM initiative, such as customer service, sales, marketing and distribution need to communicate openly with each other to assure that they support the same goals.

Timeline

CRM software installation and integration can take as little as a few weeks to more than a year depending on requirements, the existing IT ecosystem, level of customization, and the technology selected. It is also dependent upon the IT staff's skillsets and availability — both are often in short supply. This is where systems integrators may be useful and cost-effective in the long run.

As long as there is a well-conceived overall CRM strategy, which commits the company to, at some point (sooner rather than later), connecting the departmental systems to gain maximum value from its CRM initiative and investment, this implementation strategy can work successfully.

Vendor-assisted Methodology

While many CRM vendors have consulting partners that can help the implementation team to draw up the necessary re-engineered processes, it is usually much better to use in-house staff or an outside partner. The vendor is usually just a little too eager to get its product up and running to give justice to process re-engineering.

Bad Methodology

There are two bad methods of implementing CRM:

* Not mapping business or work process — CRM applications only provide business rules, not the process by which those rules are implemented.
* Taking flawed process models and trying to automate — it doesn't work — it's akin to putting lipstick on a pig.

Implementation in which the CRM technology is designed around and driven by carefully considered processes will have a higher success quotient than one that neglects process evaluation and re-engineering. While reasonable process re-engineering is a necessary pre-requisite for successful implementation of a CRM strategy, it pays to be pragmatic — find a balance between the "realistic" versus the "ideal" process. Attempting to create an "ideal" process is time consuming and, thus, costly.

Note: Integration time will be longer if the company has mainframe systems.

Moving Forward

The journey from the conference room test to final implementation involves a number of milestones with each milestone delineating clear goals and responsibilities. This is the *only* way to ensure that the I&D plan constantly moves forward. For instance:

* The CRM strategy has been determined.
* Everyone is learning to think in a customer-oriented manner.
* There's good proactive executive sponsorship and departmental and management roles and responsibilities are fully understood.
* The necessary technology has been chosen.
* The vendor(s), consultants and integrators (if needed) are on-board.
* The CRM team has dealt with the process re-engineering issues.
* Data migration is well in hand.
* Results of the system migration have been analyzed.
* Steps have been taken to remedy any problems that bubbled up during the conference room pilot test.

All *successful* CRM initiatives must have the necessary resources for fruition: high-level corporate support, a sufficient budget, experienced and dedicated in-house staff with IT buy-in, successful change management, expert project (and/or program) management, hardware, software, IT expertise, and training.

Finalizing the I&D Plan

Now is the time to wrap-up the I&D plan. Everyone should already have a good idea of the infrastructure needed to support the CRM technology, but let's review some typical requirements just to be safe:

✷ Where will the new technology (hardware and software) reside, i.e. on which server(s)? Where will the database (data storehouse) reside? What type of server(s) and databases will be used?

✷ Are additional software, development tools, middleware or hardware (e.g. database servers, ACD) necessary to correctly customize the CRM environment?

✷ Have ALL the other applications or systems with which the CRM technology must integrate been identified? There should be an up-front understanding of the impact of CRM on other corporate systems and of how the data will move between systems effectively.

✷ Have all departments and their staff members whose systems will be touched by CRM been notified of their requirements during the implementation and deployment period.

✷ Has the implementation team asked for and listened to the affected departmental personnel's feedback and taken action where necessary (and relieved fears where needed)?

✷ What about security? Is it adequate? What needs to be done to ensure there are no holes or weaknesses in the security perimeter?

✷ Is the desired functionality to data requirements mapped? Customer data is more often complex than straightforward. This usually means defining data requirements along with business requirements. At some point the implementation team will need to know what customer data is necessary and from what system it will originate. A good grasp of the level of customer data (account, demographic, psychographic, information, whatever) is critical.

✷ Is external data used (or will it be used in the future)? The use of data from external sources is accepted practice within many departments, especially marketing, to supplement the company's own customer data.

✷ Will any of the data need special preparation or cleansing, if so who is in charge of the task? Has it been done?

✷ Is the necessary process re-engineering completed or at least well underway?

✷ Who will maintain and support the CRM system once it's installed? The in-house IT department or an outside source via a maintenance and support contract?

✷ What type of workstation configuration does the CRM technology require? Does the current workstation environment support the CRM technology?

✷ Have all corporate or political barriers to rolling out the CRM been resolved? It's important to establish up-front what the tactics will be when questions of ownership or disagreements about functional priorities come up. The practical reality is that there are frequent and naturally occurring tensions within a company that can make integration, implementation and deployment planning a tortuous path. If there is any hint of a potential problem, the executive sponsor should be given a "heads up" so he or she can resolve such issues before they arise.

✱ Has the privacy policy been clearly defined and set out on paper? Regardless of whether the CRM initiative will be web-based or not, everyone *must* understand the company's boundaries for using its customers' information. CRM should not only adhere to the corporate privacy policy, it must also be an example of the company's behavior toward its customers' data.

This list is meant only as an example. Go through the CRM project's implementation and deployment process, step by step, to verify nothing has been overlooked. Identify and eliminate potential unpleasant surprises now, rather than later.

With a finalized I&D plan, implementation can begin. The implementation team's role now should be to ensure that due dates and project goals and objectives are met.

An implementation team should *never* begin by believing that since the I&D plan is pristine, the right installation team is onboard, and all the tools and middleware necessary for a smooth integration path are at hand, it'll be smooth going forward — it won't be — stuff happens. For example:

✱ Too many vendors offer minimal integration tools under the assumption that there will be only a small amount of customization needed for the CRM technology to work within a company's technical architecture. (A very unlikely scenario.) Every product requires tweaking in one way or another and most times more than "just a little" tweaking is necessary. Even implementing a single point solution will have its trying moments.

✱ With all the technology that must be melded together, no one vendor will be able to supply everything. No single vendor can provide a full range of CRM components ready to fit together like a tinker toy project. (A small point solution *may be* the exception.)

✱ The ability to create process queues, automate communication, and integrate an IVR or a web interface, all require that the implementation team and supporting staff possess a goodly amount of technical knowledge; but there will always be a "hole" somewhere.

✱ Some CRM packages incorporate their own proprietary programming language to be used for scripting internal processes. This can be problematic.

✱ Some integrators use their own homegrown tools (middleware) to speed the integration tasks along. While tempting, I would advise avoiding this approach if at all possible. Always look for off-the-shelf middleware first. If one piece of middleware doesn't get the job done, there is probably another tool that will. (The one exception is CRM in the call center environment.) But, don't let this mire down the integration process.

Note: At some point the integration team may have to trade efficiency for effectiveness and sometimes the decision must be made as to what is more important — time to deployment or everything integrated to perfection.

The term "middleware" is used throughout this section, so let's take a minute to talk about one of the most important tools in the integrator's tool box — middleware. Without middleware there is no CRM.

⊙ MIDDLEWARE

To make the system integration task a bit easier, the innovative engineering community has developed integration tools referred to as "middleware" that can make linking, for example, ERP and/or supply chain management systems to CRM systems a bit easier. There are even tools and services that can integrate systems and applications between disparate companies.

My own definition of middleware: A collection of many varieties of enabling software components that are based on widely accepted industry standards, which provide connections between and among various computing services found on disparate computing platforms and the applications that need access to those services.

There are different classes of middleware; some of which are used to integrate CRM technology. Some example are:

* *Integration brokers*, the brains of the corporate nervous system, enable businesses to manage many-to-many system interactions in an efficient manner. Just as switches improve the flow of raw data through a WAN or MAN, integration hubs powered by integration brokers improve the flow of transactions between disparate business applications.

* *Business process managers*, the memory of the corporate nervous system, enable a business to map, store, and retrieve data flows.

Middleware allows two applications to communicate with each other through a standard language.

Middleware allows applications across different platforms to communicate with each other.

word processor application   **spreadsheet application**

Windows application   **Macintosh application**

Middleware also allows legacy applications and modern applications to communicate.

Middleware takes care of transactions between servers, data conversion, authentication, and communications between computers.

Windows application   **DOS application**

Figure 32. Examples of middleware usage. Graphic courtesy of DC Team.

* *Communication middleware*, the connecting fiber, connects heterogeneous systems and maintain loosely coupled point-to-point interactions and data flow.
* *Web-integration servers*, the specialized type of middleware that enables disparate systems to communicate via Internet-based protocols are primarily used for company-to-company interactions, but they must make use of adapter, communication, or business process management middleware to bridge information flow.

CRM Needs Middleware

Middleware is vital to CRM because it can smooth out the bumps that occur due to lack of compatibility and the complexities within an IT architecture. It can easily provide many critical functions for applications and provides a firm base on which a business can integrate its disparate IT systems, intelligently linking them so they can share information. For example, middleware can:

* Enable a field service employee with a mobile device to directly connect to the CRM system and its knowledge base.
* Provide the connectivity and conduits for a SFA system to pull information from an ERP component.
* Link databases together.
* Help in the integration of data mining/data warehousing or a contact management application, sales force automation, field service, customer service/help desk and so on.

With middleware at the helm, the installation process is more open and easily upgradable, i.e. *scalable*. It lets a company mix and match software to meet its particular needs without trying to integrate proprietary packages. Also, should the

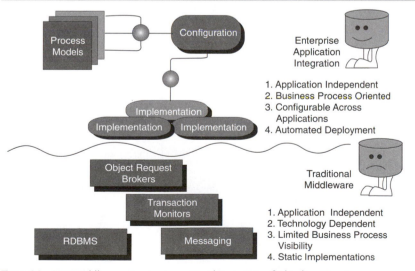

Figure 33. EAI/Middleware Comparison. Graphic courtesy of Aberdeen Group.

company merge with another, the middleware can accommodate the other enti-ty's technology until the new entity is ready to phase out one or both.

Middleware isn't the only miracle the engineers have provided to ease the pain of CRM integration. Some experts opine that the biggest miracle the engineering crowd has pulled out of their collective hat is enterprise application integration (EAI).

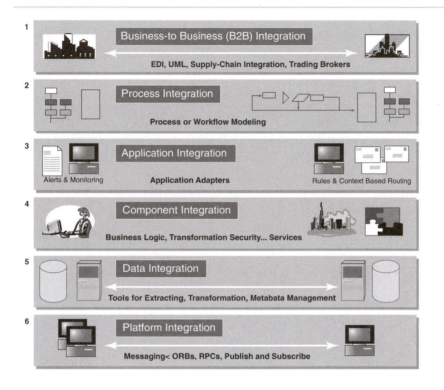

Figure 34. EAI Framework & Methodology
1. The B2B integration layer is the user or B2B interface layer and acts as the front-end for such a layered integration approach.
2. The process integration layer brings in intelligence for propagating information within the bound-ary conditions that are determined by the respective business processes.
3. The application integration layer provides a means of mapping the functions & features that have to be mapped to the individual applications.
4. The component integration layer manages the function-calls as a set of building blocks that could be applied across different applications.
5. The data integration layer performs the role of a message broker. It translates, maps and converts data in a manner such that the requests trickle up or down in the required format.
6. The platform integration layer is the carrier that enables the above integration.
Courtesy Majoris Systems Private Ltd.

Enterprise Application Integration (EAI)

EAI is a set of technologies that allows the movement and exchange of information between different applications and business processes within and between organizations. It is one way to integrate disparate systems during implementation of a CRM initiative. But as the title indicates — EAI is typically only for the enterprise-wide implementation of a full scale CRM initiative — it's an expensive undertaking.

The true benefit of EAI lies in its ability to event-enable the enterprise; i.e. limiting the amount of traditional batch processing and introducing concepts such as real-time processing and incremental processing that execute around the clock. As changes occur in one system, they are *immediately* reproduced to feed multiple similar systems. Thus the EAI approach can result in a more efficient use of resources.

EAI also can provide business process management tools to "wire" together disparate software components. However, if the company doesn't have a library of "components" (Java, C++, COM objects, etc.), there will be nothing for the tools to execute. Yet, EAI tools and technologies can provide a means for implementing the results of a business process re-engineering effort, if the business is willing to make the commitment and investment to do so.

Although their benefits may be oversold at times, EAI initiatives that are undertaken with realistic goals can actually allow some businesses to work smarter, not harder (quoting Dilbert). For example, with EAI working behind the scenes of a typical customer service event, the CRM systems can all get to the same data, even though they were written at different times on different platforms using different development technologies. Customers no longer have to give the same information every time they interact with the company.

In a nutshell, some implementation teams opt to go with an EAI solution due to the ability it gives for key systems (ERP, call center, logistics and fulfillment, e-commerce, and so forth) to be brought together as a coherent whole. This creates a cohesive IT environment while providing real-time bi-directional data transfer.

In the enterprise, an EAI-enabled implementation path can dramatically reduce time-to-benefit. For the small to mid-sized businesses, EAI is usually out of the reach of their budget. But it never hurts to investigate.

◎ AVOIDING PITFALLS

To be effective, CRM systems must be tightly integrated with the company's data repositories. Indeed, to live up to its expectations, the CRM technology must eventually be integrated with the majority of the company's applications and systems. Without such level of integration, the information contained in these disparate systems cannot be leveraged to best advantage. (It can also result in data conflicts, reports that don't make sense or, even worse, necessary data that's not available.) In other words, to work optimally, a CRM system needs to combine information stored in both front-office and back-office systems.

Back-office / Front-office

CRM is a unique combination of back- and front-office applications. If a CRM system is implemented without integration to back-office systems, it won't provide the external window to the corporate environment, meaning the value of the CRM initiative will be greatly diminished.

It's a relatively simple matter to integrate front-office systems since most share similar architecture. The real challenge is encountered in integrating back-office systems (or at least the data stored therein) and then trying to combine them with front-office systems. Back-office and front-office systems differ in their architecture — from the hardware they use to the way data is represented and stored.

Back-office systems are often old, legacy and/or proprietary technology. CRM technology is new technology, typically Unix-based client-server systems. Integration of such disparate systems into a scalable, manageable whole is the goal, but reaching that goal can cause a lot of gray hairs.

The Channel Conundrum

As if those worries are not bad enough, a company must worry about integrating not just its systems, but all the channels or contact points the customers use. Besides the normal connectivity issues, CRM technology must take into account the fact that a company's customers contact it via different modes: the Internet (web chat, callback buttons, email), traditional methods (telephone, fax, regular mail and in-person), and now, wireless. All of these touch points and their data collection streams need to be integrated into the CRM system.

Surprisingly, the failure to integrate all touch points into a system is fairly common. It occurs not because a company forgets, but because too many don't take the necessary corporate-wide view during the integration period, i.e. they don't take a holistic view and, therefore, fail to identify the impact the specific CRM initiative has on the entire customer experience.

Let's take a quick look at a CRM initiative for a call center. Some CRM teams fail to incorporate the customer information gathered by the call center's systems so it can flow into a corporate-wide data repository that is linked to other departments. For example, the marketing or sales department could use the call center's data stream in their marketing or sales campaigns. The R&D department could glean information from the call center's data banks to determine if there is a need to improve a product or even a need for a new product.

Internet

Despite all the hype, CRM and the Internet are not natural bedfellows. For example, the data held in web-based applications requires recoding so it can be used by a CRM system and vice versa. Yet, in both the bricks-and-mortar world and the cyber world accurate and up-to-date information about customers is a prerequisite for effective CRM. Indeed, the ultimate goal of CRM is to give the appearance of a perfect understanding of the needs and buying habits of each individual customer, no matter which contact channel is used.

An Amalgam

As the reader should have ascertained by now, implementing CRM isn't easy. While strategy is the leading component of a successful CRM initiative — its technology must be considered — and that technology should be an amalgam of best-of-breed technologies. Including, for instance:

* Databases, data marts and data warehouse.
* Customer service systems such as call centers and their diverse technologies.
* Sales force automation.
* Marketing automation.
* Order processing including logistics and fulfillment.
* Accounting systems.
* Field service systems.
* Web-based systems.
* ERP systems.

Eventually they all must be integrated to create a corporate-wide CRM system.

Testing

Once the implementation chores are complete, test and retest. This is the phase where the re-engineered business and work processes and CRM systems are vali-

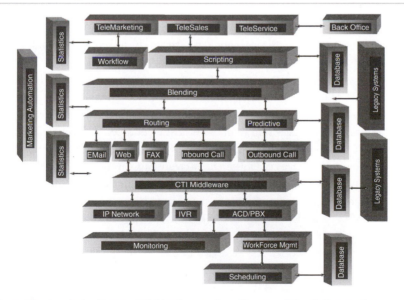

Figure 35. A typical call center CRM implementation. Courtesy of Oracle Corporation.

dated and prepared for use by authorized personnel. Once the systems works smoothly, begin user acceptance testing to review and assure that these systems and processes are sufficient to meet the requirements of the CRM initiative — the final phase prior to full-scale rollout.

Trot out the systems migration test results for review against the results of the user acceptance testing results. User testing should identify any remaining issues not caught by earlier testing efforts.

Quality Assurance

This is the phase where the new CRM technology is put through a quality assurance testing process. Review the results to determine that the new technology meets the stated acceptance criteria. Only then should the system rolled out for corporate use.

✪ SECURITY

The company should already have an on-going program of security, maintenance and monitoring to take care of potential security breaches and security holes. If not, do so IMMEDIATELY — perform a security analysis on the entire IT system and then document potential weak spots. Then perform a security audit with outside auditors to check for potential security holes that might have been overlooked. There are numerous security consultants who can help in this endeavor.

Note: *I strongly suggest that the company retain a security expert to perform a detailed review of the business's internal procedures, network topology and permissions, access controls, hardware, software and utilities that could possibly compromise the company's IT infrastructure.*

CRM Ups the Security Risk

CRM requires that disparate systems and data streams be linked into one integrated system. Today's business community tends to rely on perimeter security, which only protects the outer boundaries of an IT ecosystem. Even companies that allow employees (and trusted outside users) to access their systems via remote access use security methods that are based mainly on trust. Today, everyone from a homebound call center agent to field sales staff to management is equipped with high bandwidth connectivity from the home, dial-up access, or a wireless device that allows access to the CRM systems. While this is convenient, it creates numerous security risks.

Today's average IT ecosystem is made up of a complex combination of hardware, software and other components that are provided by a host of different vendors (hosted, custom-developed, off-the-shelf). It's virtually impossible to find and resolve all security risks. The IT staff can't really evaluate the security design of all the software components — all aren't under its charge. Therefore, as a practical matter, the IT department must depend on each component within the ecosystem being well designed from a security standpoint, and then strive to minimize exposure resulting from integration within the ecosystem.

CRM systems (and other systems as well) are continuously modified and with each change comes new security risks. This means it's almost impossible to prevent new security holes from popping up — *diligence is the only defense.*

○ FINAL DEPLOYMENT

The final deployment of the CRM initiative can begin following the satisfactory conclusion of the testing phase. That is if everything in the implementation portion of the I&D plan is completed. System architecture, network security, infrastructure, development and testing documents should all be reviewed prior to final deployment.

The implementation team or a subset of that team should authorize and facilitate the rollout stage by reviewing the completion of each component of the implementation, e.g. data imports, business and work processes, user access and security configurations.

Coordination of access for remote user devices (laptops, handheld devices, PDAs, even WAP telephones), which includes installing client components and testing synchronization, can typically be achieved with little departmental disruption. Just schedule these procedures around a conference or a sales meeting that most remote users will attend.

○ TRAINING

While training is listed as one of the last stages of the I&D Plan, it's one of the most important stages of any CRM implementation and deployment. In fact, as stated in Chapter 6, if the company doesn't provide proper training (including regular refresher courses), it can't expect to achieve maximum return on its investment in the CRM initiative.

Although the CRM vendor's generic training course may be useful, it should not be used alone — CRM is a corporate cultural change. Supplement vendor training with specifically trained "in-house trainers" or consider bringing a training consultant onboard for this stage. The training plan should incorporate not only training on the use of the actual system designed for the company, but also why the CRM technology has been put in place.

There should be four separate training groups:

* Training of the "trainers." Even if the company opts to not use its own personnel for the initial training, there will be a need for hands-on training for future hires.
* The staff (sales personnel, marketing professionals, call center agents, field service representatives, etc.)
* Departmental managers and supervisors.
* Systems administrative personnel (IT staff).

Customize the training plan to ensure that each group's training needs are met. For example:

* Sales personnel need to understand not only how to get information into the system, but also how to use the system to achieve optimal results, i.e. a higher close rate, a full pipeline, or quick identification of potential customers.
* The field service representative needs to know how to find and update information in the company's (or third party's) knowledge bank. They must understand how to upload completed work orders, download new work orders. When sales opportunities are identified, how to refer the same to the sales department.
* The call center agents need to know how to access and use the knowledge bank, answer questions on order status or account status, identify sales opportunities, make recommendations, and much more.
* Departmental managers and supervisors need to understand how to analyze the new data streams and how this additional information can help them to make their business decisions.
* The IT staff needs to learn how to provide a stable technical environment and how to enhance the system as changes or new modules are integrated, whether or not there is an outside contractor that provides maintenance and support.

At all times, the trainees should be able to access actual company data and the customized screens they will be using daily.

○ POST-IMPLEMENTATION

Once the deployment is complete, a post-implementation and deployment review must occur to ensure proper installation and system use. The post-implementation review is where the CRM team and the implementation team determine if the stated objectives of the CRM initiative have been achieved. This is also when everyone should review past mistakes so as to avoid repeating them during the next CRM project.

The chores don't end there. CRM is an ongoing process; continually evolving and must be regularly evaluated and upgraded as the business and the CRM strategy needs change over time.

Post-implementation Team

Form a post-implementation team to oversee the welfare of every CRM application after implementation and deployment. This team should be responsible for ensuring that:

* The system is easy to access
* The data provided is timely and relevant.
* Documentation and knowledge transfer is complete and satisfactory.
* New releases, upgrades and patches are installed as they become available.
* There is adequate support and maintenance.
* Security is maintained.

Documentation

Once implementation and deployment of the CRM solution is completed, the tech providers (vendors and others) should provide the company with complete and up-to-date documentation of the IT environment as it exists after the CRM systems integration is completed. (Any third party partner's contract should also provide for updates of the documentation as configurations change.) This is the time to stress to everyone the critical nature of third party knowledge transfer and to ensure that the company's IT staff understands both the process and the underlying technology of the CRM system. It's the only way the IT staff will be able to effectively support and manage the system.

This is when the underlying code used in the integration and implementation phase (referred to as the "source code") should be provided to the company for future reference. Ask a trusted programmer to review the source code to determine what has been provided and what has not been provided. Then ask the programmer to determine if the source code provided is satisfactory. Don't send final payments until the company is happy with the source code. This holds true for not only CRM vendors, but for any systems integrators brought onboard for the project.

Support

Lack of support or poor support can impact the ultimate success of the CRM initiative. Most of the support issues should have been ironed out during the tech provider's contract negotiation stage. But, to prevent needless support calls, write a support plan for the employees to follow. Make sure the support plan describes:

* Levels of support offered. First-tier (help desk, 800 number), second-tier (typically the IT department), third-tier (vendor, integrator, consultant or other third-party source).
* Hours support available for each tier.
* Procedures to be taken to escalate a support issue if not resolved through regular channels. For example, a call center agent shouldn't, except in specific circumstances, be the person who places a second-tier support call, and a call center agent should never call for third-tier support. Perhaps a supervisor or manager is the only staff authorized to request second-tier support and only the IT department is authorized to request third-tier support (third-tier support is usually quite costly).
* The procedures the IT department's staff should use to minimize support costs but at the same time efficiently deal with any problem that arises. For example, most technology providers offer telephone support as a service offering whereby a trained telephone support staff can help the *IT department* with certain issues over the telephone. Thus, the problem can be solved

quickly, without a costly in-person visit. And if the problem can't be solved over the telephone it will have at least have had the opportunity of being diagnosed prior to being escalated to an on-site visit, making the on-site call more efficient and less costly.

Maintenance

Whether handled by the in-house staff or outsourced, arrangements should be made for scheduled maintenance. Maintenance needs will vary widely from company to company. Therefore, I will just state that maintenance should include, but definitely not be limited to:

* Correctly backing up the data from all systems on a regular, scheduled basis.
* Verification that there is sufficient disk space.
* Verification that all known bugs fixes have been installed.
* Checks to verify that all applications, process and systems are running at optimal efficiency.
* Determining that the anti-virus software is current.
* A security audit has been performed in the last six months.
* Checks to verify that there are no errors recorded in the event log.

If outsourcing the maintenance and/or support service of the CRM systems, be sure the provider has the ability and capacity to undertake the full responsibility of an integrated project. The provider and its staff should have not only sufficient knowledge and experience with the specific technological infrastructure, but also experience and training in CRM ecosystems.

✪ COSTS

The expense of a successful CRM system implementation ranges widely — from as little as $1,000 all the way to $16,000 per user — of course, the CRM solutions being implemented also vary considerably.

Melding CRM within a business's operations comes with a price both in budget and effort. Let's look a generic CRM initiative and breakdown the total cost of the project.

* 27% software
* 39% integration, implementation and training
* 24% hardware
* 10% other, such as telecommunications, maintenance, service and support contracts, and so forth.

The total cost of a CRM initiative can vary from a few thousand dollars to a few million depending upon the complexity of the business and the software and hardware needed. According to a recent study by Yankee Group (a well-known research and consulting firm), a mid-sized company with an internally implemented and hosted CRM initiative for approximately 200 users (with IT infrastructure that includes a firewall, dedicated Internet access costs, VPN, back-up and recovery and security) can expect to spend close to $3 million on a CRM project. Compare those costs to an enterprise's typical CRM implementation costs, which usually run from $60 million to $130 million by the time the system is up and running to everyone's satisfaction.

IN SUMMARY

CRM implementations must be carefully considered. The short-term drawbacks of cost and implementation time must be weighed carefully against CRM's long-term benefits: higher customer retention rate, excellent customer service, targeted marketing campaigns and higher profits.

The good news is that CRM integration costs are decreasing and implementation times have shortened due to new and ever evolving CRM technology. Many businesses find that the benefits of CRM are now affordable, especially when they carefully compare their CRM options and then choose an offering that best suits their distinctive business requirements.

Some Parting Advice
Think big, but start small, and learn through doing.

As the pace of business accelerates, companies are changing more rapidly than ever. The Internet, global competition, corporate mergers, new products, new locations, new employees — all make it difficult to keep valuable customer relationships flourishing. Suddenly, companies are finding themselves working with customers that may be geographically remote as well as technologically sophisticated. CRM can help a business to prosper in such a complex environment.

The business community now operates in a new dimension of competition: namely, finding specific products and services for customers instead of finding customers for proffered products and services. This market paradigm demands that a company be thoroughly familiar with its customers and that it focuses on their needs. It's the best way a company can deploy its resources to achieve a positive outcome for everyone. CRM can help. CRM is ALL about the customer, and advances in technology have enabled many companies to exploit CRM strategies in order to build innovative, successful, customer-centric corporate cultures.

CRM is the future of business, but don't adopt it blindly. Unexpected challenges abound. As an example, a recent Alexander Group survey found that the top challenges for CRM success are not technology driven, but instead are the following (beginning with the most challenging):
* Managing employee resistance to change.
* Optimizing processes and job roles.
* Defining business requirements.
* Measuring the CRM solution's performance.

A CRM strategy must be structured with great care. It must take into account not only the company's business goals and objectives, but also its employees' abilities and needs, a realistic assessment of the products and services offered, existing technological limitations, and an intelligent CRM technology selection process. When CRM is implemented in this manner, then, with a little luck, the company will have successfully embraced CRM.

While the benefits and ROI of CRM are notoriously difficult to measure with traditional methodologies, implementing CRM in a corporate environment will profoundly affect a company's relationship with its customer base — on an individual customer basis — resulting in dramatically increased customer satisfaction and loyalty, which will bring company profitability.

My Best Advice: First, learn from others — both successes and failures. Second, always search out independent experienced advice and guidance apart from that offered by vendors. Third, continually be alert to political and cultural issues within the company and the world, in general. Hot points can bubble up at any moment and wreck havoc during any stage of the CRM initiative. Regional hot points can have far reaching repercussions for suppliers and customers. Fourth, when formalizing a CRM strategy, try to plan for any eventuality.

One Final Thought: No matter how much a company does to strengthen and deepen its customer relationships, no matter how far it goes in its transition from a product-based to a customer-based culture, there will always be more to do. There will always be new ways to make customer relationships stronger, and to do it more cost-effectively. With every step taken, the company takes another step toward using the strength of its customer relationships to lock out its competition, protect its margins, and actually outlive the products and services it probably considers its lifeblood today.

Good luck with any CRM initiative you might become involved in. It's hard, intensive work, but most businesses will benefit and grow from the good that developing a healthy customer relationship brings.

Glossary

A

ACD (Automatic Call Distributor). A switch at a call center that routes incoming calls to targets within that call center. An ACD distributes calls as workflow tasks based upon predetermined rules.

Active X. An architecture that lets a program (the Active X control) interact with other programs over a network such as the Internet.

ANI (Automatic Number Identification). A feature that provides the telephone number of the incoming caller. On some ACD systems with intelligence that can map the call number to a database, this an effect routing of the call.

API (Application Programming Interface). Software that an application uses to request and carry out lower-level services performed by a computer's operation system. In short, an API is a hook into software. An API is a set of standard software interrupts, calls and data formats that applications use to initiate contact with network services, mainframe communication programs, etc. Applications use APIs to call services that transport data across a network.

Application. A software program that does some type of task. MSWord, Netscape, Lotus Notes, CRM programs are some examples of an application.

Application to Application Integration (A2A). A form of enterprise application integration in which two or more applications are linked. The applications are usually (but not exclusively) within the same organization.

Architecture. The basic design of a system. Determines how the components work together, system capability, ability to upgrade and the ability to integrate with other systems. Refers to the overall organizational structure of a given system.

Asynchronous Communications. A non-blocking form of communication whereby applications operate independently. Thus they do not have to be running or available simultaneously. A process sends a request and may or may not wait for a response.

Autoresponders. A mail utility that automatically sends a reply to an e-mail message. They are used to send back boilerplate information on a topic without having the requester do anything more than e-mail a particular address. They are also used to send a confirmation that the message has been received.

B

B2B (Business to Business). An e-business model that refers to one business communicating with and/or buying/selling to another rather than between companies and individual customers. See B2C and E-Business.

B2C (Business to Consumer). An e-business model that refers to a business (such as a retailer) communicating with or selling to an individual consumer. B2C is the retailing part of e-commerce on the Internet. See B2B, E-Business and E-Commerce.

Backbone. A centralized high-speed network that connects smaller independent networks — typically the Internet. See Internet Backbone.

Back-end. A back-end application or program serves indirectly in support of the front-end services. This is normally accomplished through its location, i.e., closer to the required resource, or perhaps due to its capability to communicate with the required resource. A back-end application can interact directly with the front-end. However, it is more likely a program that is called from an intermediate program that mediates front-end and back-end activities.

Back-office. Any application that helps with such "back office" work as financial accounting, human resources, and manufacturing.

Bandwidth. The amount of data that can be sent through a connection. It is usually measured in bits per second.

Best-of-Breed. The best product of its type. Enterprises often purchase software from different vendors in order to obtain the best-of-breed for each application area. For example, a SFA package from one vendor and a field service automation package from another. Nobody excels in every niche.

Best-of-Class. A product considered to be superior within a certain category of hardware or software. It does not imply the absolute best overall; for example, the best-of-class in a low-priced category may be seriously inferior to the best product on the market, which could sell for ten times as much.

Bit. The smallest unit of information a computer can process and the basic unit in data communications. Bits compose a byte.

BizTalk. A set of guidelines sponsored by Microsoft, which sets out how to publish schemas in XML and how to use XML messages to integrate software programs.

Blocking Communications. A synchronous messaging process whereby the requestor of a service must wait until a response is received.

Bot. Bot is an abbreviation of the word "robot". A bot is simply a term used to describe a computer program that's utilized to perform a set of predetermined actions. The class of programs commonly known as "chat bots," are bots that interpret natural language in order to conduct semi-human conversations with users. Advances in artificial intelligence technology have made it possible for companies to employ chat bots that interact usefully with clients and customers.

Buffered Queue. A message queue that resides in memory.

Business Process. Business processes define the way in which work is done within and between organizations. They may be completely informal, rigorously structured, or anything between. A business process is an abstract set of inputs, behaviors and outputs. Each business process should deliver something of value to the business and its customers by way of one or more related activities. Business processes are often complex, and may be progressively decomposed into more detail. Processes represent the flow of work and information throughout the business. These processes act on the business entities to cause the business to function. Business processes may be long lived (such as an customer account's life cycle) or may be short lived (such as an annual report).

Business Process Management. The concept of steering work items through a multi-step process. The items are identified and tracked as they move through each step, with either specified people or applications processing the information. The process flow is determined by process logic. The applications (or processes) basically play no role in determining where the messages are sent.

Business Rules. A conceptual description of an organization's policies and practices enabling them to automate their polices and practices to increase consistency and timeliness of their business processing.

Byte. A set of bits of a specific length that represent a value, in a computer coding scheme. A byte is to a bit what a word is to a character.

C

C, C+ and C++. Very powerful programming languages which operates under Unix, MS-DOS, Windows (all flavors) and other operating systems.

Cache. To store data on a disk or in memory for quick and easy retrieval instead of retrieving it each time it is requested.

Callback. Customers visiting a website can request a telephone call from a call center agent through the use of a callback button. Information such as customer name, requested call time and telephone number are captured by the callback program.

Chat. A realtime conferencing capability between two or more users on a local network, typically the Internet. The chat is accomplished by typing on the keyboard or speaking into a microphone attached to the computer.

Chatbot. See Bot.

CISC (Complex Instruction Set Computer). A microprocessor architecture that favors robustness of the instruction set over the speed with which individual instructions are executed.

Client/Server. The client is a PC or program "served" by another networked computing device in an integrated network which provides a single system image. The server can be one or more computers with numerous storage devices. In a client/server setup the processing of an application is split between two distinct components, a "front-end" client and a "back-end" server. The client and server machines work together to accomplish the processing of the application.

Cluster. A group of computers and storage devices that function as a single system sharing one or more panel runs and working in a fault-resilient manner, allowing increased effectiveness and efficiency of security, administration and performance.

COM (Component Object Model). Microsoft's standard for distributed objects. COM is an object encapsulation technology that specifies interfaces between component objects within a single application or between applications. It separates the interface from the implementation and provides APIs for dynamically locating objects and for loading and invoking them. See DCOM.

Communications Middleware. Software that provides inter-application connectivity based on communication styles such as message queuing, ORBs and publish/subscribe.

Communications Protocol. A formally defined system for controlling the exchange of information over a network or communications channel.

Connectionless Communications. Communications that do not require a dedicated connection or session between applications.

Connectivity. The property of a network that allows dissimilar devices to communicate with each other. It also refers to a program's or device's ability to link with other programs and devices.

CORBA (Common Object Request Broker Architecture). A standard maintained by the Object Management Group (OMG).

Corporate Culture. The consistent practice of principles and values within a corporate environment.

Cross-selling. Refers to the process of increasing a customer's purchasing level by offering enhancements or new products/services, based on that customer's current purchasing status and history.

CTI (Computer Telephony Integration). Combining data with voice systems in order to enhance telephone services. Examples of CTI include automatic number identification (ANI) allows a caller's records to be retrieved from the database while the call is routed to the appropriate party; automatic telephone dialing from an address list for an outbound call; ACDs that capture call information and then pop it to a call center agent's screen.

Cyber World. The virtual world that is encompassed by the Internet.

D

Database. A set of related files that is created and managed by a database management system (DBMS) or if a relational database by a relational database management system (RDBMS). The databases that a CRM system can most effectively use are the relational databases. In non-relational systems (hierarchical, network), records in one file contain embedded pointers to the locations of records in another, such as customers to orders and vendors to purchases. These are fixed links set up ahead of time to speed up daily processing. In a relational database, relationships between files are created by comparing data, such as account numbers and names. A relational system has the flexibility to take any two or more files and generate a new file from the records that meet the matching criteria. See RDBMS and DBMS.

Database Middleware. Allows clients to invoke SQL-based services across multivendor databases. This middleware is defined by de facto standards such as ODBC, DRDA, RDA, etc.

Data Center. A centralized storage facility typically used by a hosted solution provider or other service provider to retain database information for remote access by end users.

Data Level Integration. A form of EAI that integrates different data stores to allow the sharing of information among applications. It requires that the data be loaded directly into the database via its native interface. It doesn't involve the changing of business logic.

Data Mart. A data storehouse that covers a single subject — essentially a separate database fed by operational systems and built solely to warehouse all the data that is collected to support a specific CRM initiative. Data marts contain data tailored to support the specific analytical requirements of a given department or business function that utilizes a common view of the data. Thus providing more flexibility, control and responsibility. Data marts may be set up to perform simple querying and reporting, to enable trending and multidimensional analysis, to support sophisticated data mining such as pattern recognition, or to allow unstructured exploration of the data. See Data Warehouse and Database.

Data Mining. A method for exploring detailed business transactions. As the term implies, it refers to digging through gigabytes of data to uncover patterns and relationships contained within the business activity and history. Although data mining can be done manually by slicing and dicing the data until a pattern becomes obvious, normally specialized software is used to analyze the data automatically.

Data Mining Software. Applications that can discover meaningful correlation, patterns, and trends by shifting through large amounts of data stored in repositories, using pattern recognition technology as well as statistical and mathematical techniques.

Data Transformation. A key requirement of EAI and message brokers. For example syntactic translation is used to change one data set into another (such as different date or number formats), while semantic transformation is used to change data based on the underlying data definitions or meaning.

Data Warehouse. A data storehouse that serves as a single repository of data and usually replaces multiple data marts. Once a data warehouse is loaded with cleansed data, it provides all the data feed and receives all the data streams for the entire enterprise. Then the appropriate query, analysis tools and data mining software are integrated to allow a company to parse its data for better decision making. See Data Marts.

DBMS (Database Management System). A collection of applications that enable the storage, modification and extraction of information from a database. There are many different types of DBMSs. They can range from small systems that run on personal computers to huge systems that run on mainframes. Also DBMSs have differing characteristics, for example, terms such as relational, network, flat, and hierarchical all refer to the way a DBMS can organize information internally. The internal organization affects how quickly and flexibly the information within the database can be extracted.

DCOM (Distributed Component Object Model). A Microsoft protocol that enables software components to communicate directly over a network. DCOM is based on the DCE-RPC specification and works with both Java applets and ActiveX components through its use of Microsoft's COM.

Directory Services. A network service that identifies all resources on a network and makes them accessible to authorized users and applications.

Distributed Computing Environment (DCE). A suite of technology services developed by the Open Software Foundation for creating distributed applications that run on different platforms such as RPC, distributed naming service, time synchronization service, distributed file system and network security.

DML (Data Manipulation Language). SQL statements that can be used either interactively or within programming language source code to access and retrieve data stored in a database management system.

DOM (Document Object Model). A platform- and language-neutral interface that will allow programs and scripts to dynamically access and update the content, structure and style of documents.

DSI (Decision Support Interface). DSIs typically consist of easy to use analytical tools utilized to distill information from data receptacles, such as data warehouses and data marts.

Due Diligence. A comprehensive investigation and assessment of all attributes, issues and variables inherent in a target entity/person/product/service that will impact the target's ability to achieve its strategic objectives.

E

EAI. See Enterprise Application Integration.

E-Business. The use of Internet technology and advanced networking to extend and enhance the traditional business model. Buying and/or selling electronically over a telecommunications system. See E-Commerce, B2B and B2C.

E-Commerce. Using the Internet and e-business technology to enable a business to conduct commerce electronically. See E-Business, B2B and B2C.

Encryption. A system of using encoding algorithms to construct an overall mechanism for sharing sensitive data. The translation of data into a secret code.

Enterprise Application Integration (EAI). A set of technologies that allows the movement and exchange of information between different applications and business processes within and between organizations.

ERP (Enterprise Resource Planning). A business management system that integrates all aspects of a business, such as, product planning, manufacturing, purchasing, inventory, sales, and marketing. ERP is generally supported by multi-module application software that helps to manage the system and interact with suppliers, customer service, and shippers, etc.

Execution. In marketing or sales the actual delivery of an outbound message to a segment or target group over a specific delivery channel.

Extranet. A private, TCP/IP-based network that allows qualified users from the outside to access an internal network. A type of intranet that allows authorized outsiders to access specific areas of the network, except that extranets differ from intranets in that an intranet resides behind a firewall and is accessible only to people who are members of the same company or organization, whereas an extranet provides various levels of accessibility to outsiders. See Intranet and VPN.

F

Firewall. Hardware and/or software that sit between two networks, such as an internal network and an Internet service provider. It protects the network by refusing access by unauthorized users. It can even block messages to specific recipients outside the network.

Front-end. (1) A program that interfaces with and services the initial user. (2) A front-end application is an application that users interact with directly. (3) Relative to the client/server-computing model, the client part of the program is often called the front end and the server part is called the back-end.

Front-office. Usually refers to an application designed to help with the management customer-facing activities, such as sales, marketing, and customer support.

FTP (File Transfer Protocol and File Transfer Program). Allows users to quickly transfer files to and from a distant or local computer, list directors, delete and rename files on the distant computer. FTP the program is a MS-DOS program that enables transfers over the Internet between two computers.

G

Gateway. A combination of hardware and software that performs translations between disparate protocols so that there can be communication between different types of networks. In data networks, gateways are typically a node that connects two otherwise incompatible networks and often perform code and protocol conversion processes.

Groupware. A class of software that's a model for client/server computing based on five foundation technologies: multimedia document management, workflow, e-mail, conferencing and scheduling. Groupware is typically used to help groups of colleagues (workgroups) attached to a local-area network organize their activities.

H

Hacker. An unauthorized person who breaks into a computer system to steal or corrupt data.

Hardware. Objects that go with the computing environment that can be touched. For example, modems, interface cards, floppy disks, hard drives, monitors, keyboards, printers, motherboards, memory chips, etc.

HTML (Hypertext Markup Language). A set of markup symbols inserted in a file intended for display on a web browser. The markup instructs the browser how to display a web page.

I

IIOP (Internet Inter-ORB Protocol). A protocol developed by the Object Management Group (OMG) to implement CORBA solutions over the World Wide Web. Unlike HTTP (which only supports transmission of text), IIOP enables the exchange of integers, arrays, and more complex objects. IIOP ensures interoperability for objects in a multi-vendor Object Request Broker (ORB) environment.

Information Technology (IT). All aspects of managing and processing information, especially within an enterprise. Can also be known as Information Services (IS) and Management Information Services (MIS).

Infrastructure. The interconnecting hardware and software that supports the flow and processing of information.

Integration. (1) A combination of units so that they work together or form a whole. (2) A process in which separately produced components or subsystems are combined and any problems due to the interaction are addressed. (3)

Activities by which specialists bring different manufacturers' products together so as to form a smoothly working system. (4) Products or components that are integrated and appear to share a common purpose or set of objectives or they observe the same standard or set of standard protocols, or share a mediating capability. (5) Products or components that were designed at the same time with a unifying purpose and/or architecture although the individual units may be sold separately even though they were designed with the same larger objectives and/or architecture, share some of the same programming code (such as special knowledge of code).

Integrator. See Systems Integrator.

Internet. A public global network of computers that exchange data.

Internet Address. A registered IP address assigned by the InterNIC Registration Service.

Internet Backbone. The worldwide structure of cables, routers and gateways that form a super-fast network. It is provided by number of ISPs that use high-speed connections linked at specific interconnection points (National Access Points referred to as NAPs).

Intranet. An internal TCP/IP-based network behind a firewall that allows only users within a specific enterprise to access it.

IP Address. A unique identification consisting of a series of four numbers between 0 and 255, with each number separated by a period, for a computer or network device on a TCP/IP network.

IT See Information Technology.

IVR (Interactive Voice Response). An automated telephone answering system that responds with a voice menu and allows the user to make choices and enter information via the keypad. IVR systems are widely used in call centers as well as a replacement for human switchboard operators. The system may also integrate database access and fax response.

J

Java. A high-level object-oriented programming language similar to C++ from Sun Microsystems designed primarily for writing software to leave on Web sites which is often downloadable over the Internet. Java is basically a new virtual machine and interpretive dynamic language and environment.

JDBC (Java DataBase Connectivity). A Java API that enables Java programs to execute SQL statements similar to ODBC.

K

Keyword. In database management, a keyword is an index entry that identifies a specific record or document. In programming, keywords (sometimes called reserved names) can be commands or parameters, which are reserved by a program because they have special meaning.

L

LAN (Local Area Network). A short distance data communications network consisting of both hardware and software and typically residing inside one building or between buildings adjacent each other — thus allowing all networked devices to share each other's resources.

Legacy Systems. Existing information resources, programs and systems currently available to an enterprise. Legacy systems are operational mainframes, personal computers, serial terminals, networks, databases, operating systems, applications, and other forms of hardware and software that have great value to an enterprise. However, these legacy systems cannot be easily or economically extended to the web or otherwise modified to fulfill the requirements of an e-business model.

Load Balancing. Automatic balancing of requests among replicated servers to ensure that no server is overloaded.

M

Mail List. A program that allows a discussion group based on the e-mail system.

MAN (Metropolitan Area Network). A communications network that covers a geographic area such as a city or suburb. See LAN and WAN.

Message Broker. A key component of EAI. An intelligent intermediary that directs the flow of messages between applications. Message brokers provide a very flexible communications backbone and provide such services as data transformation, message routing and message warehousing.

Message-Oriented Middleware (MOM). A set of products that connects applications running on different systems by sending and receiving application data as messages. Examples are RPC and message queuing.

Message Queuing. A form of communication between programs. Application data is combined with a header that contains information about the data to form a message. The messages are then stored in queues, which can be buffered or persistent. It is an asynchronous communications style and provides a loosely coupled exchange across multiple operating systems. See Buffered Queue and Persistent Queue.

Message Routing. An application process where messages are routed to applications based on business rules. Therefore, a particular message may be directed based on either its subject or actual content.

Message Warehousing. A central repository for temporarily storing messages for analysis or transmission.

Meta Data. Data about data. Meta data is essential for understanding information stored in data warehouses. Meta data describes how and when and by whom a particular set of data was collected, and how the data is formatted.

Middleware. Software that facilitates the communication between two applications. It provides an API through which applications invoke services and it controls the transmission of the data exchange over the network. Middleware serves as the glue between two applications. Middleware can also be referred to as "the plumbing" because it connects two sides of an application and passes data between them. See Communications Middleware, Database Middleware, Object Middleware and Systems Middleware.

N

Non-Blocking Communications. An asynchronous messaging process whereby the requestor of a service does not have to wait until a response is received from another application.

O

Object Management Group (OMG). A consortium of object vendors whose goal is to provide a common framework for developing applications using object-oriented programming techniques. It is the founder of the CORBA standard.

ODBC (Open Database Connectivity). A Microsoft Windows standard API for SQL communication. A standard database access method that allows databases such as dBASE, Microsoft Access, FoxPro and Oracle to be accessed by a common interface independent of the database file format.

ODBMS - See OODBMS.

ODS (Operational Data Store). A collection of data used to support the tactical decision-making process. It's subject-oriented, fully integrated, holds current data and is volatile.

OLAP (OnLine Analytical Processing). A relational database system capable of handling complex queries through the use of multidimensional access to the data (viewing the data by several different criteria), intensive calculation capability, and specialized indexing techniques.

OLTP (OnLine Transactional Processing). Processing transactions as they are received by the computer, which means that master files are updated as soon as transactions are entered at terminals or received over communications lines.

OMG. See Object Management Group.

OODBMS (Object-Oriented DataBase Management System aka ODBMS). A database management system (a program that lets one or more users simultaneously create and access data in a database) that supports the modeling and creation of data as an object.

Open Applications Group (OAG). A consortium formed to promote the easy and cost-effective integration of key business application software components.

ORB (Object Request Broker). Software that allows objects to dynamically discover each other and interact across machines, operating systems and networks.

Order Processing. All of the activities associated with filling customer orders - computer related and human related.

Outsourcing. Contracting with outside tech specialists such as consultants and system integrators, or perhaps a software vendor, service bureau or solutions providers.

P

PC (Personal Computer). A computer designed for use by a single person versus simultaneous use by more than one person.

PDA (Personal Digital Assistant). A handheld device that uses a stylus for input and combines computing, telephone/Internet and networking features.

Persistent Queue. A message queue that resides on a permanent device, such as a disk, and can be recovered in case of system failure.

Personalization. Refers to technology that enables the determination of a customer's interest based on his or her preferences or behavior, constructing business rules to decide how to deal with such a person, and dealing with that person according to those preferences.

Pop-up Window. A second browser window that "pops up" when called by a link, a button or an action.

Portal. A website that offers a broad array of resources and services. These resources and services can be offered to the general public, to a specific authorized group, and/or maintained for an enterprise's internal organization.

Protocol. A set of rules governing the format of messages that are exchanged between computers and people.

Publish/Subscribe. A style of inter-application communications. Publishers are able to broadcast data to a community of information users or subscribers, which have issued the type of information they wish to receive (normally defining topics or subjects of interest). An application or user can be both a publisher and subscriber.

Q

Query. A request for information from a database. There are three general methods for posing a query (a) Choosing parameters from a menu wherein the database system presents a list of parameters from which a choice can be made. Although this is the easiest method to use when posing a query, it is also the least flexible. (b) Query by example (QBE) wherein the database system presents a blank record and lets the user specify the fields and values that define the query. (c) Query language wherein the database systems requires the user to make requests for information in the form of a stylized query that must be written in a special query language. Although this is the most complex method, but it is also the most powerful.

Queue. Lined up awaiting something. Customers and customer queries are frequently placed in queues awaiting specific service resources, such as a call center agent.

R

RDBMS (Relational Database Management System). A database management system that stores data in the form of related tables referred to as relational databases. RDBMS enables one or more people to simultaneously create, update and administer a relational database. RDBMSs are powerful because they require few assumptions about how data is related or how it's extracted from a database. As a result, the same database can be viewed in many different ways.

Real-Time. Occurring immediately. The data is processed the moment it enters a computer, as opposed to BATCH processing, where the information enters the system, is stored and is operated on at a later time.

Remote Procedure Call (RPC). A form of application-to-application communication that uses a tightly coupled synchronous process, which hides the intricacies of the network through the use of an ordinary procedure call mechanism.

Response Measurement. The ability to track the response rate of a marketing message or sales offer that's delivered to a specific target, whether as a group or on an individual basis.

Response Modeling. The ability to develop response models within a marketing automation tool to be used in future segmentation and targeting efforts.

RFP (Request for Proposal). A document that invites a tech provider or vendor to submit a bid for hardware, software and/or services. It may contain generalized information or consist of a document containing very detailed specifications.

ROI (Return on Investment). How much "return," usually profit or cost saving, results from a particular action. ROI calculations are sometimes used along with other approaches to present a business case for a given project.

Router. A device that connects any number of LANs. Routers check the destination address of the packets and decide the route to send them. Very little filtering of data is done and routers don't care about the type of data they handle.

S

Scalable. Refers to the ability of a hardware or software system to adapt to changing conditions, especially an increase in demand on its resources.

Segmentation. The process of identifying groups of customers around which to conduct sales and marketing efforts by analyzing the existing customer base.

Server. A computer or software package that provides a specific capabilities to client software running on other computers. For example, a Web server has a very fast permanent connection to the Internet and subsystems to protect against power outages, hackers and system crashes. A database server manages and processes the database and database queries.

Service Level Agreement (SLA). A contract between a provider and an end user which stipulates and commits the provider to a required level of service. A SLA should contain a specified level of service, support options, enforcement or penalty provisions for services not provided, a guaranteed level of system performance as it relates to downtime or uptime, a specified level of customer support and what software and hardware will be provided and for what fee.

SKU. See Stocking Keeping Unit.

SLA. See Service Level Agreement.

Sockets. (1) A software object that connects an application to a network protocol. (2) A portable standard for network application providers on TCP/IP networks. (3) A receptacle into which a plug can be inserted, such as for a microprocessor or other hardware component.

Software. Computer instructions or data - anything that can be stored electronically is software.

SQL (Structured Query Language). Pronounced "sequel", it is a standardized database query language.

Stocking Keeping Unit (SKU). A method of identifying a product without using a full description.

Stored Procedure. A program that creates a named collection of SQL or other procedural statements and logic that is compiled, verified and stored in a server database.

Supply Chain. A group of physical entities such as manufacturing plants, distribution centers, conveyances, retail outlets, people and information which are linked together through processes (such as procurement or logistics) in an integrated fashion, to supply goods or services from source through consumption.

Supply Chain Management (SCM). A set of skills and disciplines, including those of IT, which shepherd a product from its original design to its ultimate delivery to the buyer.

Synchronous Communications. A form of communication that requires applications to run concurrently. A process issues a call and until it receives a response.

Systems Integrator. A specialist that brings different vendors' products together so as to form a smoothly working digital information system.

Systems Middleware. Software that provides value-add services as well as inter-program communications. An example is transaction processing monitors which are required to control local resources and also cooperate with other resource managers to access non-local resources.

T

TCO. See Total Cost of Ownership.

TCP (Transmission Control Protocol). A transport layer, connection-oriented, end-to-end protocol that provides reliable, sequenced and unduplicated delivery of bytes to a remote or local user.

TCP/IP (Transmission Control Protocol/Internet Protocol). A networking protocol (the Internet's protocol) that provides communication across interconnected networks, between computers with diverse hardware architectures and various operating systems. It runs on virtually every operating system. IP is the network layer and TCP is the transport layer.

Total Cost of Ownership (TCO). A model that helps businesses to understand and manage the budgeted and unbudgeted costs incurred for acquiring, maintaining and using an application or a computing system. TCO normally includes training, upgrades, and administration, as well as the purchase price.

Trigger. A stored procedure that is automatically invoked on the basis of data-related events.

Two-Phase Commit. A mechanism to synchronize updates on different machines or platforms so that they all fail or all succeed together. The decision to commit is centralized, but each participant has the right to veto. This is a key process in real time transaction-based environments.

U

Up-selling. Refers to the process of increasing a customer's purchasing level by encouraging the customer to switch from one product to another through the offering of additional recommendations when the customer is browsing and/or placing an order.

V

Value Chain. The chain of all the companies involved in developing or delivering a particular product or solution, from raw-material supplier through final retailer and sometimes the end-user.

Vendor. (1) The seller. (2) A firm or individual that supplies goods or services including software companies and hardware manufacturers in the computer industry.

Vertical. A vendor or provider that offers products and/or services that covers the needs of a specific vertical market or industry, such as: telecommunications, health care, banking, manufacturing, education, retail, real estate, government, law, steel, durable goods, automobile manufacturing or computers.

VPN (Virtual Private Network). A private network that is configured within a public network. VPNs enjoy the security of a private network via access control and encryption, while taking advantage of the economies of scale and built-in management facilities of large public networks.

W

WAN (Wide Area Network). A network that is geographically scattered with a broader structure than a LAN. It can be privately owned or leased, but the term usually implies public networks.

WAP (Wireless Application Protocol). A secure specification that allows end users to access information instantly via handheld wireless devices such as mobile phones, pagers, two-way radios, etc.

Web. A subset of the Internet that in today's world is accessed via a Web browser.

Workflow. The automatic routing of documents to the users responsible for working on them. (1) Workflow provides the information required to support each step of the business cycle. The documents may be physically moved over the network or maintained in a single database with the appropriate users given access to the data at the required times. Triggers can be implemented in the system to alert managers when operations are overdue. Automating workflow sets timers that ensure that documents move along at a prescribed pace and that the appropriate person processes them in the correct order. (2) In marketing workflow is the process of managing and coordinating the activities of a marketing campaign from planning and budgeting to execution and tracking.

Work Process. Work process is the automation of procedures where documents, information or tasks are passed between participants, according to a defined set of rules to achieve or contribute to an overall business goal.

X Y Z

XML (eXtensible Markup Language). An open standard used for defining data elements on a web page and business-to-business documents. It uses a similar tag structure as HTML; however, whereas HTML defines how elements are displayed, XML defines what those elements contain. HTML uses predefined tags, but XML allows tags to be defined by the developer of the page. Thus, virtually any data items, such as product, sales rep and amount due, can be identified, allowing web pages to function like database records. Thus, by providing a common method for identifying data, XML can support business-to-business transactions.

Zero Latency. No delay between an event and its response.

Zero Latency Enterprise. An enterprise in which all parts of the organization can respond to events as they occur elsewhere in the organization, using an integrated IT infrastructure that can immediately exchange information across technical and organizational boundaries.

Zero Latency Process. An automated process with no time delays (i.e. no manual re-entry of data) at the interfaces of different information systems.

Index of Figures

1. The Importance of Business Objectives in the Deployment of CRM Initiatives. 004
2. Components of CRM. 007
3. The Customer Pyramid. 009
4. Profit Impact of 5% Increase in Retention. 015
5. 80/20/30 Rule of Customer Profitability. 021
6. Key CRM Initiatives. 022
7. Economic Benefits of CRM. 028
8. Components of a Company-wide CRM Strategy. 031
9. Customers are not Created Equal. 032
10. Customer Relationship Management: Building a New Infrastructure. 036
11. Profit Impact of Retention. 041
12. Benefits that a Company can attain from a Loyal Customer Base. 068
13. The CRM Ecosystem. 094
14. How CRM Anaysis Works within a Typical Marketing Department. 102
15. A Multidimensional Database. 108
16. CRM Closed Loop Process. 112
17. How Key Stages of a Customer's
    Life Cycle can be Affected by Isolated Departmental Silos. 122
18. OLAP can Transform how a Sales Organization
    creates and distributes information for better decision making. 127
19. Sales Analysis is Key to Successful Sales Management. 133
20. Sales Analysis Tools Can Help. 134
21. How Do Salespeople Judge their SFA Systems? 136
22. Average Costs to Service Customers - Per Contact Channel. 144
23. Multidimensional Analysis. 150
24. Trend Analysis. 151
25. The CTI/CRM Partnership. 153
26. Wireless, CRM, Call Centers and Field Service. 163
27. A Typical Enterprise CRM System's Data Environment. 172
28. A Typical Data Warehouse System. 174
29. A Typical CRM System and Its Data Environment. 175
30. A Basic Process Map. 248
31. Example of a Purchasing Process Map. 249
32. Examples of Middleware Usage. 254
33. EAI/Middleware Comparison. 255
34. EAI Framework & Methodology. 256
35. A Typical Call Center CRM Implementation. 259